# PRINCIPLES OF MILITARY MOVEMENTS
## CHIEFLY APPLIED TO INFANTRY

### ILLUSTRATED BY
## MANOEUVRES of the PRUSSIAN TROOPS

#### OUTLINE OF THE BRITISH CAMPAIGNS IN GERMANY DURING THE WAR OF 1757

COLONEL DAVID DUNDAS

The Naval & Military Press Ltd

published in association with

Published by
**The Naval & Military Press Ltd**
Unit 10 Ridgewood Industrial Park,
Uckfield, East Sussex,
TN22 5QE England
Tel: +44 (0) 1825 749494
Fax: +44 (0) 1825 765701
www.naval-military-press.com

*in association with*

ROYAL
ARMOURIES

*In reprinting in facsimile from the original, any imperfections are inevitably reproduced and the quality may fall short of modern type and cartographic standards.*

Printed and bound by Antony Rowe Ltd, Eastbourne

# PRINCIPLES

OF

# MILITARY MOVEMENTS,

CHIEFLY APPLIED TO INFANTRY.

ILLUSTRATED

By MANOEUVRES of the PRUSSIAN Troops,

AND BY AN

OUTLINE of the BRITISH CAMPAIGNS in GERMANY, during the WAR of 1757.

TOGETHER WITH

An APPENDIX, containing a practical ABSTRACT of the Whole.

BY COLONEL DAVID DUNDAS.

LONDON;

# To the KING;

THIS WORK,

HUMBLY OFFERED IN THE HOPE OF ITS

PROVING IN SOME DEGREE

USEFUL TO HIS MAJESTY'S SERVICE,

IS,

WITH THE MOST PROFOUND RESPECT,

INSCRIBED,

BY HIS MAJESTY'S

DUTIFUL

SUBJECT AND SERVANT,

DAVID DUNDAS.

# PREFACE.

THE want of uniformity, of method, and of regulation, in the movements of the BRITISH troops has been long felt, the attendant bad consequences too apparent, and the remedy earnestly and universally desired.

There may have been times when an army in these countries was esteemed dangerous to liberty, but it is now considered as indispensably necessary to retain empire, and has become so interwoven with the constitution, support of government and the welfare of the whole, that although its apparent existence is only annual, it may be looked upon as permanent, ought to be cherished, and its improvement should be cultivated to the highest extent—From the low establishment of this army in time of peace, we labour under every disadvantage at the commencement of a war. Besides well founded apprehensions for home, our foreign possessions are at such time in the most imminent danger from an enterprizing enemy, for we then are but ill enabled to detach a considerable body of formed troops to their assistance: ships alone cannot protect them, nor can ships alone attempt any considerable conquest.

On the policy or necessity of such a situation when the rest of Europe remains in an armed state, we presume not to decide—But the lower the ordinary numbers of our army are, the higher ought to be the discipline which inspires it, the more exact should be the system of uniformity which directs it, and the deeper ought to be the foundations laid, for its immediate increase and quick training when war breaks out. Under these ideas are the MILITIA of England justly included, on whose numbers and spirit their country has so full and so deserved a reliance—they have no time to employ in attending to the showy and less essential parts of parade, which have been so much cultivated. To them should the most simple, most essential, and most universal modes be minutely pointed out, of performing that service which the very beginning of a war may require from them.

---

Considerable improvements have taken place in the general tactique within these few years; and while the other nations of Europe have endeavoured to copy the lessons of the great masters, most of the alterations that we have made are not for the better; more founded on particular fancy, and caprice, than on the uniform, comprehensive, and immutable principles which should ever direct the operations of great bodies, and of their component parts. Our soldiers are active and brave, our officers intelligent and willing, much good internal œconomy exists in our infantry; many of our battalions are individually well drilled, but no system regulates or connects their joint exertions. During the late war, the annual encampments formed in the different parts of England offered the most favourable opportunity of establishing uniformity of discipline in all its branches. It is only to be regretted that no such attempt was made; that a proper direction was not given to much laudable zeal, and good disposition; and that we concluded as we began, leaving every one in a great measure to follow his own mode and imagination.

---

After

( iii )

After a service of many years in varied situations, of much regimental practice, and of five campaigns in the war of 1760—a thorough conviction of the want, and of the necessity of a permanent and general system of discipline in the British army has now induced the writer to pursue in part, so desireable an object: he has only to lament his inability to execute it in a manner more clear and satisfactory. He has endeavoured to avail himself of information wherever he could find it, and flatters himself that he has not been unsuccessful in pointing out many of the principles and reasons on which the Prussian practice is founded. In this he has been more particularly instructed and assisted by the invaluable work of General Von Saldern, and from the opportunities he had in 1785 of attending the Prussian exercises, which have enabled him to revise and correct what he had long before more imperfectly arranged. In a work of this kind repetition is unavoidable, often necessary, and should be excused; much inaccuracy, he relies, will not be found. His language he presumes not to defend, and trusts it will not be made the object of criticism; if it suffices to convey clear and distinct ideas of his subject, its other defects he hopes will be overlooked.

---

The formation of the officer and soldier is his professed design: the preparation of the materials with which the general is to act his great object: the elements of movement, and the circumstances connected with them, have principally claimed his attention. The absolute necessity of uniformity, and unity of system he has endeavoured to incultate and demonstrate. He has avoided as much as possible going beyond his prescribed line, or entering on the higher scenes of service. He presumes not to deal out lessons to the general, but at the same time must remark, and means to enforce that a practical and ready knowledge of the detail, and execution of all possible movements of the troops, ought to be the first and leading study of all ranks, and is indispensable in the general as well as in the soldier; otherwise his application of them to situation and circumstances

cumstances will be accidental and inaccurate.—Our service in time of peace affords too few opportunities of extensive practice; in time of war it is dangerous, and too late to make experiments; our regulations should therefore be the more pointed, explicit, and prepared for that event.

---

The general system hereafter laid down is as far as the writer's knowledge extends, taken from that followed by the great armies of Europe, and of its excellence there can be no doubt. This attempt at its explanation may perhaps appear prolix; but in a work of this kind, where it is wished to convince as well as to lay down precept, an assent to its propriety can only flow from a full possession of the principles on which it is founded. If it possesses no other merit, it has however that of the most ardent zeal, and earnest desire to promote the honour, and ensure the superiority of His Majesty's arms. But should it also in the smallest degree tend—to put an end to the uncertain practices of our service; the prevalence of review show; of loose desultory movements, and consequent unconnected support; of want of solidity, &c. &c.: to convince officers, that partial modes are incompatible with that general uniformity which alone invigorates and gives energy to the collective body: to recall and fix the attention of the troops to the great and immutable principles that should ever direct their movements; and should it be found to point out in any degree those principles and their practical application, the warmest wishes of the writer will be most amply gratified.

# CONTENTS.

## INTRODUCTION.

| | Page |
|---|---|
| REVIVAL of military science | 1 |
| Modern improvements | 2 |
| Former and present State of { Cavalry | 4 |
| Artillery | 6 |
| Infantry | 7 |
| Prussian principles and system | 8 |
| Defects of British practice | 11 |
| Light troops { Foreign / British } | 13 |
| Situation of the general officer | 15 |
| Advantages of an established system | 17 |
| General heads | 18 |

## PRINCIPLES of MOVEMENT.

| | Page |
|---|---|
| General principles | 21 |
| Movements made on fixed points | 22 |
| Application of movements | 23 |
| Words of command | 27 |
| Signals | 28 |
| Music, drums, &c. | 29 |

| | Page |
|---|---|
| POINTS of { March | 29 |
| Formation | 30 |
| Appui | 31 |

## Of the ALIGNEMENT.

| | Page |
|---|---|
| Points fixed in the alignement | 32 |
| Intermediate points determined | 33 |
| Points ascertained } by adjutants / Line prolonged | 34 |

## INSTRUCTION of OFFICER and RECRUIT.

| | Page |
|---|---|
| Officer | 36 |
| Recruit | 37 |

General

|                                         | Page |
|---|---|
| General attentions of instructor        | 38 |
| Position                                | 39 |
| Of the march in front                   | 40 |
| ———— march in file                      | 44 |
| ———— oblique march                      | 45 |
| ———— side march                         | 46 |
| ———— halting, dressing                  | 47 |
| ———— wheeling                           | 48 |
| Distance of { files / ranks }           | 50 |

### Of the COMPANY.

| | |
|---|---|
| Its composition and formation | 56 |

### Of the BATTALION.

| | |
|---|---|
| Its composition and formation | 59 |
| Mode of instruction | 65 |
| Attentions in exercise | 66 |
| Attentions of general officers at reviews, &c. | 67 |

### Of the COLUMN. 69

### Of the OPEN COLUMN.

| | |
|---|---|
| General circumstances attending the open column | 70 |

### The BATTALION in OPEN COLUMN.

| | |
|---|---|
| When the battalion breaks to a flank | 75 |
| Entry and march in the alignement | 76 |
| Halt, formation in line, and dressing | 79 |
| When the rear of the column files to enter the alignement | 83 |
| When the column forms line on a { rear / central } alignement | 84 |
| Countermarch of the column { Of divisions singly / Of the wings / Of the battalion on the center } | 86 / 87 / 89 |
| Taking distances from the rear | ib. |

### The LINE in OPEN COLUMN.

| | |
|---|---|
| General march in column | 91 |
| March in the alignement | 92 |
| Formation of the line from column | 93 |
| Arrival of a considerable column in the alignement | ib. |
| Entry of the rear battalions of the column into the alignement when the head halts | 94 |
| Arrival of several columns in the alignement | 95 |
| March and formation of Prussian infantry | ib. |

### CHANGES of POSITION in the LINE.

| | |
|---|---|
| General principles | 96 |
| General mode of changing position | 97 |
| The line breaks to the directing division. General situation of the open column to form in line | 98 |
| Relative positions { parallel / intersecting } | 99 |
| Taking a new position on a central division | 100 |
| Forming when the central division is moveable | 101 |
| General attentions in change of position | 104 |
| Change of position by the diagonal march of divisions | 105 |
| Situations which require the exclusive movements of the open column | 106 |
| Necessity of wheeling divisions backwards into column | ib. |
| Wheel in the column of march | 107 |

### Of the CLOSE COLUMN.

| | |
|---|---|
| General circumstances attending the close column | 109 |
| Close column when { line / open column } formed from | 111 |
| Forming the battalion from close column on the { Front / Rear / Central } division of the column | 112 |

|  | Page |
|---|---|
| Forming the line from a considerable close column | 116 |
| Formations of two columns into line to the {front / flanks / rear} of the march | 118 |
| Formation in two lines of several columns | } 119 |
| Formation from several columns | |
| Prussian movements in 1785 | 121 |

### MOVEMENTS of the BATTALION in front.

|  | Page |
|---|---|
| When to march | 127 |
| ———— halt | 129 |
| ———— dress | ib. |
| ———— incline | 130 |
| ———— alter the direction of the march | } 131 |
| ———— retire | |
| ———— front | |
| Pass an obstacle | ib. |
| When the battalion wheels on a flank backward or forward | 133 |
| ———————— Passage of lines | 134 |
| Reforming the first line {parallel / oblique} | 135 |
| Refusing a wing | ib. |

### MOVEMENTS of the LINE in front.

|  | Page |
|---|---|
| When to march in front | 136 |
| ———— halt / dress | 140 |
| ———— retire | 141 |
| ———— front / pass obstacles / incline | 142 |
| Prussian movements in line | 143 |

### Of the ECHELLON.

|  | Page |
|---|---|
| General application and movements of the echellon for the battalion and line | 144 |
| Attack in echellon from the {parallel line / oblique line / column} | 152 |
| Movement in line and echellon | 155 |
| Prussian movements in echellon | ib. |

### Of SECOND LINES.

| | Page |
|---|---|
| General rules in the movements of second lines | 156 |
| Change of position of second lines | 160 |

### PASSAGE of LINES.

| | Page |
|---|---|
| Passage of the first line through the second | 164 |
| Reforming the first line | 166 |
| Throwing back a wing after passing | 168 |
| Retiring a line and marching on the enemy's flank | 170 |
| March and oblique formation on an enemy's flank | 171 |
| Inversion in line, and Prussian movement | 172 |

### Of the RETREAT.

| | Page |
|---|---|
| General method of retreat | 176 |
| Of the chequered retreat in line | 177 |
| Refusing a wing in retreat / Chequered retreat of two lines | 178 |
| Retreat of infantry before cavalry {in line without baggage / with baggage} | 180 |
| Manœuvre at Berlin in the Square | 183 |

## Of the DEFILE.

Passage of the defile { to the {front / rear} of the column } { from the {flank / center} of the line } — 185
General attentions in passing a defile — 187
Reducing / Increasing } the front of the march — 188

## COLUMN of MARCH.

General principles — 189
Various attentions necessary in the march — 191
——— March by lines to the flank } — 193
——— March to the front or rear }
Columns of march formed to {front / rear / flank} — 194
March and attack of small corps posted — 195
Marches of an army in different situations — 196
March and formation of an army previous to the attack — 197
Necessity of regulation for the marches of an army — ib.

Regulations given for the marches and movements of the British and French armies in Germany during the war of 1756 — 198

OUTLINE of the British campaigns in Germany — from 207 to 263

## APPENDIX.

ABSTRACT of words of command and circumstances of execution necessary in most of the preceding movements.

Form of review, and movements necessary for the practice of a single battalion, and of a small corps.

Idea of a manœuvre for a small corps.

# INTRODUCTION.

THE revival of military science may be dated from the days of the NASSAUS, and GUSTAVUS ADOLPHUS, who in their arms and tactique, blending the Grecian, and Roman systems, formed armies eminent in discipline, and renowned for the great actions they atchieved.

But at that time the slender revenues of princes only brought moderate numbers into the field: those not always well appointed, and often entirely dependant on the country in which they made a predatory war for a precarious subsistence: there were few places of great strength: armies were too small to carry on a war of positions: a battle decided the fate of a large tract of country: the vanquished retired to a distance to recruit: the victors at their leisure carried on their subsequent operations in cantonments, and did not again unite, till the approach of the enemy called them anew to the field: heavy armed cavalry still made the principal force of the army: the greatest proportion of the infantry were armed with pikes; the rest with matchlocks, which, as well as the artillery of those times, were very despicable.

*Revival of military science.*

Since that period the improvement in fire arms, and the arts of modern fortification; the alterations in the conftitutions of moft countries in Europe; the numerous bodies of men, which the ambition of Louis the XIVth brought into the field; the univerfal eftablifhment of ftanding armies, and the modern methods of conducting war, have made a great change in the military art, and increafed the difficulty and intricacy of it as a fcience.

After the experience of above a century, during which time each arm and fyftem has had its warm partifans: the pike has given way to the improved mufquet and bayonet, as to a weapon more univerfal, and adapted to all varieties of fituation.

<small>Modern improvements.</small>

The thin order of formation (bringing more men into action where fire is now fo material) has become the fundamental one in preference to that on a greater depth, which however is ftill retained, and applied on particular occafions.

Defenfive heavy armour has been laid afide as infufficient againft mufketry, and productive of more inconvenience than advantage.

The lance is alfo difufed, as requiring a man to be covered with heavy armour, and as incompatible with the clofenefs and vigour of the charge in which the ftrength of cavalry is allowed to confift.

Artillery has been increafed, and the ufe of it improved in a furprifing manner.

The internal difcipline and œconomy of the troops have been well eftablifhed, and the authority of the officers fully founded: nor is there a modern inftance of great mutiny or defection; which were fo common, when corps and armies were the property of individuals, or of the general, rather than of the ftate or prince whom they ferved.

If the compofition of modern armies is inferior as to the fpecies of men; the manner of modelling them fufficiently compenfates; where the authority of the prince, of the general, of the officer, and the fubordination of each to each is fo thoroughly enfured.

The

The laſt great improvements in the military art we owe to the KING of PRUSSIA; who (maſter of the ſcience, and practice of the profeſſion) to the moſt perfect uſe of their arms, that men are capable of, and to the obedience, and order of the Grecian ſyſtem, has added the Roman art of movement and manœuvre, which before him, our numerous, and extended armies, but ill underſtood. *Pruſſian ſuperiority of ſyſtem.*

From the ſuperior diſcipline of his troops, and from the wonderful reſources of his own genius, he has atchieved actions not to be paralleled in antiquity; events in which fortune has had little ſhare; not combats of Greeks and Romans againſt Barbarians, where the advantages of the former were ſo deciſive: but battles of Germans againſt Germans, armed and nearly diſciplined alike—His extraordinary talents alone turned the ſcale.

It is rivality, and a parity of attack and defence, that beget improvement. As all Europeans are armed in the ſame manner, and mean to follow the ſame military ſyſtem, it is in the more dextrous uſe of their arms, and in a more profound knowledge of the ſcience of war, that one nation muſt hope to have a decided advantage over another. *Reaſons of improvement.*

From a near equality we muſt expect exertion, not from decided ſuperiority. We are not to look for it in the *Eaſt*, where that of the Europeans, (and thoſe not the beſt conſtituted troops) is at leaſt as remarkable as in the days of the ancients.

There can be no inſtance more ſtriking of the improvement of modern military ſcience, than the late and conſtant ſucceſs of the RUSSIAN againſt the TURKISH troops. *Turkiſh defects.*

Theſe laſt, from the nature of their religion, government, and manners, cannot be much degenerated. There was a time, and even then European bravery was diſtinguiſhed, when they were at leaſt equal to their neighbours, and threatned the conqueſt of chriſtendom: but not keeping pace with the improvements of laſt century,

century, and servilely adhering to their old customs; we have seen, by degrees, the Germans successfully make head against them; and the Russians, at last, rise so superior, as with a very inconsiderable force to endanger the capital of their empire, which without the interposition of other powers, would certainly have fallen.

Overawed by the success and discipline of the Russian troops, we have seen an immense army (brave as individuals) refuse to follow their general, and ignominiously desert that standard, under which their ancestors had conquered so many kingdoms.

---

Ancient state of cavalry.

When the strength of armies consisted in CAVALRY, the lance men charged in a single rank, and then retired in order to renew the charge, or to allow the cuirassiers (who advanced at no great pace) to improve the advantage with their swords. The *melée* and confusion became general; the exploits of individuals were wonderful, and determined the fortune of the day, more than the uniform and united shock of large bodies, and those repeated and supported.

But when a higher degree of order came to be introduced into armies, the German cavalry, who in their Turkish wars were opposed to a more numerous and nimble enemy, were obliged to trust to their armour, and to the solidity of their order, and became only capable of slow movements, advancing to the charge at a gentle trot.

The French cavalry, from the impetuosity of the nation, were accustomed to charge with velocity, though generally disunited and in great disorder; yet this was sufficient in the time of Turenne and Luxembourg, to give them a decided superiority over the solemn German cavalry, whom we find again successful, when animated by the superior genius of Marlborough and Eugene.

---

But no certain principles were yet established, and the slow regularity introduced into the cavalry, tended to cramp the spirit, instead of adding to the vigour of the body; every thing seemed calculated

calculated for resistance only; a squadron was thought hardly capable of a trot; for a line it was supposed impossible; ranks and files were so crouded, that the least violent movement, put them in confusion: It remained a lifeless and inactive mass, till put in motion by the wonderful talents of the KING of PRUSSIA.

<small>Modern cavalry improvements.</small>

He saw that the great business of the cavalry, is the offensive, rather than the defensive; giving the attack, rather than receiving it; that the velocity of its movements, and quick change of place, must ensure the most decided advantage over an enemy, inferior in either, and while it inspirits the one body, in the same proportion tends to intimidate the other.

He has shown the facility of moving great bodies of cavalry from one point to another.

From experience, he has proved the possibility of a full line arriving upon the enemy with rapidity and order; and in the impulse of the charge has shown, that the spur tends as much to overset the opposite enemy, as does the sword which should complete that defeat.

Sensible, that the great intervals betwixt squadrons when in line, only tended to make them irresolute, and to swerve in advancing upon the enemy, under pretence of taking them in flank; he formed his first line of cavalry in a full line, and insisting on the rapidity of the charge, obliged them to the direct attack upon the opposite enemy. Behind that line he had his bodies of reserve to support; to repair any loss; or improve any advantage gained.

The superior advantages of the PRUSSIAN system, has obliged most of the nations of Europe to adopt it, and to endeavour to improve upon it.

The French since the war of 1756, sensible of their want of method and order, have taken uncommon pains. But perhaps by attending more to the minute than the essential, they have not hitherto made all the progress they might have done. Still the alteration for the better is very evident in their cavalry; and they are now capable of a vigorous and united charge, instead of that

<small>Cavalry.</small>

*en fourageurs:* a term, implying the disorder in which they formerly attacked.

---

The improvements of ARTILLERY have kept pace with those of the other arms; and it now makes a principal part of every military establishment. From being unweildy, unmanageable, and uncertain in its effects; it is become numerous, light, and transportable—accompanying the movements of the troops, worked with dexterity, and often turns the scale of victory.

Although it certainly confines and retards the progress of armies; yet there are situations where it decides, and none where the advantages that attend it can be given up. Its combinations with the movements of the infantry and cavalry, are therefore an essential part of the General's calculation; and every thing that tends to accelerate its motions, becomes a most desirable circumstance.

*Artillery.*

No place can now long resist the improved modes of attack, and the heavy and unremitting fire of a numerous well directed artillery. In the field, where battles are often reduced to affairs of post, its operations are of the greatest consequence: and where an enfilading fire in a greater or less degree, can be obtained (and which is always endeavoured at) its effects are great, and it prepares the way for a less vigorous resistance from the enemy.

For some time past it has been the custom, to attach one or more field pieces to each battalion. But this method has by many been thought disadvantageous; and that it is better to assemble them in brigade; and to post them properly according to circumstances, rather than to trust to their accidental positions on the flanks of battalions.

In all armies therefore, the conduct and management of the artillery is a very great part of military combination, and influences much the movements and operations of the whole.

---

If the improvement in the tactique of the cavalry and artillery have been considerable; that of the INFANTRY has not made less progress. Independent of internal œconomy, and arrangement, and
the

the moſt perfect uſe of their arms that men can attain; the king of Pruſſia has renewed and perfected the ſcience of movement and manœuvre, which in latter times was but little underſtood.

From the time that artillery became formidable, the muſquet the univerſal weapon of the infantry, and that the thin order of formation took place; our modern and extended armies had become unmanageable. Former ſtate of infantry.

Though frequent, great, and perſevering marches had been occaſionally performed, yet each was the product of intricate arrangement, calculated for one particular purpoſe, and no other. Hours were taken up in forming in line of battle in the proceſſional manner, which was then the only one known: and when ſuch line was once formed, it was difficult to make any conſiderable alteration in it, without much previous explanation, and endangering the order of the whole.——Attacks were planned upon ſome determined ſituation, in which it was expected to find the enemy, and the columns of march were compoſed accordingly: but on approaching him, if he had taken a different poſition, it was not eaſy to adapt the attack to ſuch change of circumſtance; and the precious and deciſive moment of action was often loſt in explaining the new and neceſſary diſpoſition.

The infantry were brave and ſteady, though ill trained—their movements were ſlow—their formations tedious, and not varied to apply to all ſituations—the relative motions of the line were but ill underſtood, and worſe executed on all ſides: and moderate generals were much embarraſſed in the conduct of a machine ſo complicated, and where the great ſprings of action were ſo ill regulated.

It was reſerved for the KING of PRUSSIA to diſcover in the mechaniſm and application of the marching and cloſe *Column*, a ſimple remedy for all theſe defects, and to apply its principles to the training of all bodies great or ſmall. The juſtneſs of them being once aſcertained, he has enſured the preciſion and ſameneſs of execution by poſitive rules, from which no one dare vary, and by the unremitting attention of his ſervice.——This ſyſtem, founded by his predeceſſor, improved by the experience of his camps of peace, and ſanctified by Pruſſian improvements.

his

( 8 )

his great fuccefs in war, has upheld the wonderful fuperiority of his arms for thefe forty years; and as yet he is but followed at a diftance (not overtaken) by the reft of the world.

Speculation only, aided by the practice of fmall bodies, will often miflead, when applied to the movements of large ones. It is in the camps of peace alone where juft rules can be determined. In time of war, there is feldom opportunity or fufficient exactnefs of execution, to afcertain the neceffary and minute parts of fuch a fyftem. In camps of peace therefore, the theory has been eftablifhed, and fuccefsful wars have confirmed the practice.

---

Simplicity and generality of principle are the diftinguifhing features of his fyftem—by the precifion to which every thing is reduced; by the permanent and great divifions of the troops in the field, which he firft eftablifhed, he has organized his army in fuch a manner, that in his hands, the moft numerous has become as flexible and manageable as the fmalleft body; nor has he ever failed in the face of the enemy to perform movements, that have always been decifive in his favour.

*Pruffian principles.* The formation of his columns are fimple, in direct progreffion, and from either, but always the fame flank of the different bodies which compofe them. Their unfolding and formation into line on any one given divifion, as circumftances require a greater or leffer number of men to be thrown to either hand, are equally fimple, and never made on accidental points, but always on fuch as are fixed, determined, and relative to the fituation of an enemy.

The precifion with which his troops, when in line, can incline to a flank, at the fame time that they advance in front, enables them on many occafions imperceptibly to outflank an enemy.

The movements in *echellon* give a great facility in advancing: the corps better fupport in the attack, and in the retreat more effectually protect by gradually falling back upon each other.

At the moment of attack, his ranks, files, and different bodies compofing the lines are clofe; and fhould his enemy, lefs firm and

compact,

compact, present an opening, it is there he seizes the opportunity of entering with his cavalry, and probably gaining a decided advantage.

Quick movements, with considerable columns or lines of infantry, he considers as impracticable and ruinous from the hurry and disorder that must thence ensue, and which a good cavalry will in an instant take advantage of: but brigades, or smaller divisions of the line, occasionally lengthen their step, and move on with rapidity at the moment of attack.—Although the general movements of the infantry may appear slow and solemn, yet they are so accurate, that no unnecessary time being lost in dressing, or correcting distances, they arrive sooner at their object than any other, immediately form, and at the same instant proceed in perfect order to the attack.

<small>Prussian principles.</small>

Superior order, regularity, and weight of fire, are the great objects of the Prussian infantry, and the rapid movements of cavalry, make this circumspection necessary. His cavalry are bold, active, and enterprizing, and never fail to seize and improve any advantage that offers; their order, rapidity, and quick movements, are equalled by no other, and have been often decisive in the day of battle.

The discipline of his troops, ensures success against an enemy not capable of such rapid movements, who cannot suddenly counteract him, and with difficulty know where he will direct his force, as his dispositions change in an instant, and his formations are made under cover of any advantage the ground presents, as well as of the movements of cavalry, and of the smoke and fire of a numerous artillery, which make it less necessary to sacrifice infantry in attacks of post, soon dislodge an enemy, and reduce the decision to opener ground.

Organized as this army is, a few general intimations, instead of long and detailed orders, are sufficient to put it in motion for any purpose of march or attack.—The general, the officer, and the soldier, are each capable to execute his part, and each has the mode of movement invariably ascertained to him, whatever possible change

Prussian principles.

of situation can be required; and from which mode no individual dare deviate.

---

Thus prepared, and moving with such facility, he seldom waits the attack.—When he approaches the enemy, it is at the head of a strong advanced guard, behind which his columns are collected and combined, according to circumstances.—In that situation, he determines on his future disposition; the heads of his columns are carried to their several given points; distances are just; the celerity of forming is great; the detail of execution, in all possible cases, is known to every one.—One part of his army is strengthned by reserves, or the advanced guard; the rest of his line is refused: the flank of the enemy is probably gained; and wherever he makes his attack, by these means he acts with superiority, although on the whole, his army may be inferior in number.

Though his combinations, as a general, were infinite; yet the requisites in the soldier were few, though essential, and never varied; otherwise, from the disadvantages of composition, his army could not have maintained, (considering its yearly losses and recruits) that capacity and *superiority of manœuvre, so conspicuous,* during the whole course of the last war.

The countries in which his armies have acted, offered all the varieties of wood, mountain, plain, and inclosure: the same principles of movement and formation have still taken place, modified according to situation: they apply equally to his infantry, cavalry, and artillery, singly, or when combined: the smallest portions of his troops, are individually trained to act their part in the columns or lines of the army.

---

Permanent and detailed *regulations* for the conduct of every military individual, in every possible situation, obviously useful in all services, are particularly necessary in the BRITISH. Such regulations prepare the materials, wind up the springs, and give unity and energy to the whole machine: without such, chance and

( 11 )

and caprice direct, negligence and confusion follow, and the operations of so Tartar like an army, can only be attended with ruin and disgrace.

It is our misfortune to have had no line of conduct laid down: the good order of regiments has less depended on the rules of the service, than on the accidental efforts of individuals, and on the fashion of the day, equally as that changing: where zeal or science were wanting, the consequences have been too often apparent. Necessity of regulation.

---

Hence our very thin and extended order to make more show—an affected extreme of quickness on all occasions—the running of one movement into another, without those necessary pauses which tend to show their propriety, and justness of execution—the system of central dressing, filing, and forming on almost all occasions—the single person attempting to direct the battalion, and its parts, in every situation, in order to beget a false and improper precision—the forming and breaking on the move, the easier to conceal and cover lost distances and accidental lines, which otherwise would be apparent—the several methods of wheeling established—the different and false composition of columns, which each battalion at pleasure adopts—the chance movement of the line in front, regulated by no fixed principle. Defects of the British practice.

Unaccustomed to form or move on determined points, (the necessity of which, is not so immediately striking in the management of a single battalion) a given position is taken up with no degree of precision—filing, which was formerly little known or practised in the infantry, is now general, and often misapplied where division-marching should take place—the hurry practised by individual regiments, becomes improper and impossible when acting in conjunction with others; confusion and inaccuracy follow; and time is consumed in endeavouring in vain to correct those errors, which original method would prevent from ever arising.

Our ranks are so thin, our files so open, and such intervals permitted between companies of the same battalion when in line; that all idea of solidity seems lost.—We have begun at the wrong

**Defects of the British practice.**

end, and have endeavoured to deduce the useful movements of the line, from the showy and review ones of the battalion. We have established as general rules, what ought to be regarded only as exceptions—we have started from parade, as our primary order, instead of considering it as only secondary to that of attack—our conclusions have been false, as our *data* were defective.

Nor do these irregularities operate only in the field, and in great bodies. They equally take place in the internal composition and management of our battalions; each has its singular mode of discipline, unknown to the other, and often as opposite as those of two distinct services.—A detachment or guard, is as heterogeneous a body as an army, and the command of it on a small scale, is attended with the same difficulty. The whole forms a scene of intricacy, which no individual can be sufficiently master of, and which nothing but the substitution of one over-ruling and universal method can ever dissolve.

---

The very small proportion of *cavalry* employed in the American wars, has much tended to introduce the present loose and irregular system of our infantry.—Had they seen and been accustomed to the rapid movements of a good cavalry, they would have felt the necessity of more substantial order, of moving with concert and circumspection, and of being at every instant in a situation to form and repel a vigorous attack.

**Reasons of those defects.**

The weakness of our battalions in time of peace has farther confirmed it; for not sufficiently restrained by regulation, an attention to show has universally taken place, and the facility of dancing about a small body has deluded many, who if they had larger numbers to direct, would have found the necessity of method, and of more order in their operations.

In other services, garrisons being numerous, and regiments composed of two battalions at least, they are in the practice of exercising, and manœuvring together; and therefore such modes will always be prescribed as are general, and proper for combined movements in the field: but our single battalions not feeling that necessity,

necessity, from seldom acting with others, and having no proper system laid down for them, naturally adopt such partial methods as seem to each convenient, and may perhaps be so for a single battalion, although totally repugnant to, and destructive of the unity of the line when several are to act in concert.

The importance also which the *light infantry* have acquired, has more particularly tended to establish this practice. During the late war, their service was conspicuous, and their gallantry and exertions have met with merited applause.—But instead of being considered as an accessory to the battalion, they have become the principal feature of our army, and have almost put grenadiers out of fashion. The showy exercise, the airy dress, the independent modes which they have adopted, have caught the minds of young officers, and made them imagine that these ought to be general and exclusive.—The battalions, constantly drained of their best men, have been taught to undervalue themselves, almost to forget, that on their steadiness and efforts, the decision of events depends; and that light infantry—yagers—marksmen—riflemen, &c. &c. vanish before the solid movements of the line. *Light infantry.*

In all the armies of Europe, there is a great proportion of light infantry and cavalry; but they do not unnecessarily deviate from the general principles of the service, nor are their peculiar modes adopted by the more important bodies of the line.

They form separate corps, but still preserve the greatest order. Their skirmishers and dispersed men are loose, detached and numerous, according to circumstances; but a firm reserve always remains to rally upon, and to give support as may be wanted— their attacks are connected, and their movements the same as the rest of the line—their great province is to form advanced and rear guards; to patrole, to gain intelligence, to occupy the out posts, to keep up communications, and by their vigilance and activity to cover the front, and ensure the tranquillity of the army—they *Foreign light troops.*

decide

decide not, nor are they chiefly relied upon in battle, although on many such occasions they perform regular and eminent service.

---

Whether the establishment of our battalion light companies, is an advantageous mode, may admit of some doubt. When assembled in corps, they ought to act as other battalions do; but while attached to their several regiments, they had best be considered as out of the line, and placed in the rear of the battalion as a reserve, ready to sally forth, and execute the part allotted them.

<small>British light infantry.</small>

There seems no reason why the light infantry should not conform to the same principles of order and movement, as the battalion. The frequent dispersion and peculiarities which they are taught, should be considered as occasional exceptions. By their present open order, and independent ideas, they are under very little controul of their officers; and their practice seems founded on a supposition of the spirit and exertion of each individual, more than on the real feelings by which the multitude are actuated. Were our battalions, also, more accustomed to act in line, and with cannon, they would see the impropriety of every instant scattering and throwing forward the light infantry, whose situation must often prevent the proper use of the artillery.

Our present prevailing modes, are certainly not calculated either to attack or repulse a determined enemy, but only to annoy a timid and irregular one—they are not general, but were first adopted in local situations that may not soon recur. There is great danger in an *irregular* system, becoming the *established* one of a British army; and the most fatal consequences may one day ensue, if we do not return to a due sense of the necessity of solidity, effort, and mutual dependance, which it is the great business of discipline to inculcate and regulate.

---

In the *progress* of military science, the other nations of Europe have of late revised, simplified, and improved their system of discipline: but we have rather complicated ours.—For the modes of our battalions being various as their numbers, each is wedded to its own partial practice; and the whole form an aggregate of parts, too obviously discordant when assembled together, and when any conjunct and regular operation is required from them. Their different methods must be first reconciled, or a neutral one substituted; each is tenacious of its own; much arrangement is required, and in this manner we are always beginning, always changing, and never arrive at the object we have in view.

Consequences of the want of system.

The fatal consequences arising from this want of system and concert, have been severely felt, and the remedy of universal regulation long demanded, as the only means that can keep in any kind of order, the distant parts of an army, which has never the advantage of assembling in large bodies in camps of peace, as is the practice of other nations, or of being formed into considerable divisions, in cantonments and quarters, under the permanent inspection of the general officer, but where on the contrary, even the companies of the same battalion are separated from each other the greater part of the year.

Necessity of establishing of it.

---

The situation of the GENERAL OFFICER in our service, is at present, particularly embarrassing. Let his talents be ever so eminent, with such jarring materials as our battalions now present, it is absolutely impossible for him to attempt the most simple manœuvres before an enemy, much less such complicated ones, as the circumstances of his situation might point out and require. He can take no advantage; he can risque no movement; and the moment which is so precious in all military operations, is lost in previous arrangement, at such an instant ill explained and worse understood. Instead of finding his part, and that of every individual, ascertained by the general rules of the service, he is either obliged to give way to chance, and risk universal confusion, or to endeavour to procure unity and concert, by substituting his own temporary and perhaps crude methods.

Peculiar situation of the general officer.

But were system and science established and strictly adhered to; there would only remain for him, as is his great and distinguishing province,

province, to judge and determine of the application, and propriety of movements with refpect to the enemy's fituation. He would not then be embarraffed in explaining the manner of execution, which in all poffible cafes fhould be known to each individual, fhould be invariable, and follow as foon as he fignifies his intentions.———Famed as we are for the valour of our foldiers, what might not be expected from the ability and zeal of our generals, if the fame means of inftruction and conduct, which regulates other fervices, exifted in ours!

It is needlefs to collect reafons for the promulgation of a general *military code*. The experience of every hour points out the immediate neceffity of it; and that it is high time that uniform method fhould take place of accidental caprice, variety, and fafhion.

<small>Neceffity of a military code.</small>

There are many proper and excellent regulations and cuftoms, exifting in the Britifh fervice; but it is difficult to know which are obfolete, and which are in force. They are no where collected under one view; are filent on fome of the moft material heads of difcipline, and fall very fhort of laying out a line of conduct for each individual. They have been framed at various times, to remedy the inconveniences of the day; and do not originate from the one great and fundamental point, to which all military regulations, as to a center, fhould refer.

<small>Execution of it.</small>

However numerous and comprehenfive the different articles neceffary to military inftruction may appear, yet the detailing and filling up the moft important of them is perhaps not difficult to attain. Many of the materials exift in our own, many are to be found in other fervices: arrangement and the labour of competent individuals will accomplifh the reft; and, although the parts are dependant on each other, yet they may with propriety be feparately promulgated.

But in the progrefs of fuch an undertaking, it is neceffary to afcertain the great *principle*, which fhould be conftantly held in view,

view, and to the completion of which, each circumstance should tend. This principle we must look for in the *field* operations of the troops; it must necessarily refer to their actions and exertions; and can only have for it's object, the conquest of the enemy by superior movement and arrangement. All circumstances of show, discipline, regulation, &c. &c. must therefore tend to this point, and be subservient to it.

<small>Great leading principles.</small>

The ultimate object of the army being fixed, we must follow it through its several divisions; and in tracing the principles of its just movements in the line, the column, the *echellon*, &c. we shall see the means by which its great end is accomplished. In its intermediate place we shall find the battalion, and discover the necessary detail and application of these circumstances to its various operations. We must determine its primary and most essential order, that in which it meets an enemy; every other change of its situation must be relative thereto, and its component parts formed accordingly. Regulations of discipline, police, &c. must constantly be referred to and examined by this test, which should be ever kept in view, and nothing suffered to interfere with, or militate against it.

From this broad basis, we must ascend in regular gradation, till we see the single soldier stationed on the point of the pyramid—we must reason from the great to the less; and after ascertaining the effects, discover the means and the causes.

---

Such a general system when introduced, should be enforced, and upheld by the strong hand of authority; and no deviation permitted from it, at the discretion of individuals; nor should any one have a right to dispense with observances, which so nearly concern the general good, and which by situation he may be placed to watch over. Whatever changes time and circumstances might make necessary, should come from authority—when established and well understood, the military machine and all its parts would then with ease be managed; every thing be in unison; its efforts consolidated and formidable, and the minds and knowledge of individuals

<small>Advantages of system when once established.</small>

dividuals be enlarged—nor would this tend to confine or cramp genius, but rather to regulate and direct it aright; and full scope would be still left for military talents, in the application of just principles, to the boundless variety which time, numbers, situation of ground, and other circumstances will always present.

Positive regulation should ever speak in a tone of command, and not make use of unnecessary explanation. But such tone is not here assumed; and therefore, besides laying down rules, it is wished to discover, and point out the substantial reasons of each, in order to induce conviction.

---

In pursuit therefore of this object, it is here proposed to enter on one great division of such a work, and to establish a system of order and movement, founded on the most acknowledged and undoubted principles, and by which the *field service* of the troops, as far as it relates to those circumstances, should be regulated.

---

The GENERAL HEADS, under which it will appear, are as follows.——The subdivisions which spring from them, are in their place detailed.

*Principles of Movement* — { General, and such as should on all occasions direct the various operations of the troops.

*Points of* { *March*, *Formation*, *Appui*, *Alignement* } The use of these in order to procure accurate and just movements, are explained.

*Instruction of Officer and Recruit* { The essential and fundamental rules of training, are here pointed out; but a fuller detail of several of the articles will be necessary.

*Company*
*Battalion* { Their composition and formation, modes of assembling, instruction, and exercise.

| | |
|---|---|
| Movements { Battalion } in Open<br>of the { Line } Column | All possible formations from column, are made by one general method, which requires no previous directions being circulated, is adapted to all varieties of surface, and allows the formation of the line to a certainty, at any one instant, or in any position, however intricate. |
| Changes of Position of the Line | Changes of the line from one situation to another, in embarrassed ground, are made upon the principles of the open column. As filing is the great mean used in all these movements, the flanks which in that case seem presented to the enemy, need give no uneasiness, as their relative situations, and the precautions taken at every moment, will ensure them from harm. An instant is sufficient to form in order of battle, and to put the body, however great, in a posture of defence against the enemy's cavalry, if they attempt to interrupt the movement. |
| The close Column - - - | |
| March in Front of the { Battalion<br>Line | Rules are given for the march of great bodies in front, which is a most essential and difficult operation. |
| The Echellon - - - - | Shows its general principles, and the application of its movements to the battalion, column, or line. |
| Second Line - - - - | Shows the methods of executing its relative movements, in consequence of any change of position made by its first line. |

| | |
|---|---|
| *Passage of a First Line through the Second* - - - - | When pressed by an enemy and obliged to retire, and take up a new position in the rear. |
| *Passage of the Defile* - - | By the line or column, and to front or rear. |
| *The Retreat* - - - - | The general conduct of the retreat of lines, and the defence that infantry can make, when retiring in line or columns against cavalry. |
| *Column of March* - - - | General rules, which ought to regulate the marches of all bodies, are given and exemplified in the practice of the British and French armies in Germany. |
| *General circumstances essential to be observed in the movements of Battalions, Lines, Columns, &c.* - - - | These circumstances are briefly recapitulated and pointed out. |
| *Abstract of words of command, and circumstances of execution necessary in the movements, and formations of the Open Column, Close Column, Battalion, Line, &c.* | Points out the particular nature of each movement—the separate words of command, whether given by leaders of battalions or divisions, and the order in which they follow. As also the chief circumstances of execution attending each—the whole being meant as a practical abstract of what has been more particularly detailed. |
| *General heads of review exercise for a Battalion or larger Body* - - - - - | Deduced from the established principles, meant to comprehend the most essential movements, and given as an example of what may be required of a battalion or larger body, at a regular review or inspection. |

The application of the principles laid down under these several heads, embrace all the general and particular operations of an army, and may be modified and adapted to all situations, and in all varieties of ground. They are occasionally exemplified in the combinations and movements of considerable bodies, and in the actual exercises of the PRUSSIAN troops.

A knowledge of these paths necessarily conducts to the more elevated and sublime points of the military science; and to eminent rank, distinguished talents, and great experience any further or deeper investigation would belong. But the many capital and instructive works on the higher parts of war, joined to the illustrious examples of able generals, sufficiently elucidating such scenes, furnish complete lessons for those who ought to profit from them, and leave us only to regret, that the inferior but essential branches of the profession have not been equally cultivated and well understood.

## PRINCIPLES of MOVEMENT.

THE manœuvres of the battalion and of smaller bodies must be deduced from those of the line, calculated for all varieties of ground; and such only adopted as are of use in the field, and equally applicable to one or to fifty battalions.

All partial methods therefore of movement or formation fitting only for a battalion or its parts should be rejected—Such as are not connected with, or cannot be considered as those of the great body, are defective, useless, and may be even hurtful, as tending to destroy the unity and harmony of the whole.

The duties of the commander of line, battalion, company, division, as well as of every other individual, must be detailed in such a manner; that any number of battalions when assembled, may at once execute any manœuvre required, by the same one universal and determined method.

*General principles.*

Complication in exercise, or movement must be avoided: every possible operation of the battalion must be adapted to the capacity of the soldier, or at least of the non-commissioned officer, who may frequently have his part in the command.—Where intelligence can be united with great practice and attention, perfection in the execution will be soon attained.

---

As the movements of lines, or of great bodies, should only be the application, and repetition of those of the battalion—the method of taking up any new position should therefore be fixed, and obvious to every individual.

*Movements made on a fixed point.* All movements must be made on determined distant points, which the commanders of leading bodies will scrupulously observe: uncertainty and wavering will be thereby avoided in the march; the fatigue, and attention of the soldier lessened; the direction of the body always regulated according to circumstances; and the greatest nicety, and ease obtained in the formation of the line.

*Simplicity in the execution.* The soldier must not be puzzled with various and unprofitable ways of attaining the same end.—Simplicity and generality of principle are the great objects to be aimed at; they imply perfection in the military machine, as much as in the mechanical one.

In the execution of manœuvres, the greatest precision is essential and requisite in small bodies, as the want of it will operate in proportion to the extent of large bodies.

---

*Cavalry and infantry principles the same.* The principles for cavalry and infantry must be as analogous as the nature of the two arms will admit—The same words of command, and methods of formation and movement should as far as possible be adopted, and take place in their separate or combined manœuvres. The great advantages thence arising are sufficiently obvious.

*Close order the fundamental one.* The order of attack, or close order, must always be considered as the primary, and fundamental one; from which those of exercise and parade are deduced and occasionally taken.

( 23 )

No movement is in itself (properly speaking) dangerous; it is only so with regard to its application, and the calculation of its being finished before advantage can be taken of it.

No movement should be begun, till the commander is satisfied with the propriety and safety of its execution. And whatever troops can perform judicious movements with the greatest celerity and order, will always be enabled to place themselves in the most advantageous situation to give or receive the attack; or if circumstances require, to avoid it. *Application of movements.*

The practice of movement in all kind of ground is what alone can give the requisite *coup d'œil*, which in an instant sees defects, and remedies them; seizes advantages, and improves them; calculates justly the time necessary in the execution, according to local circumstances; above all, enables at once to take a decided part, which never fails to communicate vigour to those that are to execute—In all cases indecision is fatal. *Coup d'œil.*

The intelligence and readiness of officers is the great spring of all action and movement; unless that point is attained, it is in vain to expect exertion from the soldier. The regulations of the service should rigidly insist upon it—From the principles which ought to direct the conduct of officers, one would suppose it no matter of difficulty; yet general experience evinces the necessity of the most express rules for that purpose: none so important as that the deserving should always be distinguished from the negligent and incapable.

An officer who cannot thoroughly discipline and exercise the body intrusted to his command, is not fit in time of service to lead it to the enemy; he cannot be cool and collected in the time of danger; he cannot profit of circumstances, from an inability to direct others; the fate of many depends on his ill or well acquitting himself of this duty. It is not sufficient to advance with bravery; it is requisite to have that degree of intelligence, which should distinguish every officer according to his station: nor will soldiers ever act with spirit and *Officer.*

anima-

animation, when they have no reliance on the capacity of those that do conduct them.

In military plans and arrangements, those that execute are not so sanguine as he who designs; and therefore the event generally falls short of the expectation: experience should make allowances, and not require more than human nature in common furnishes.—What we have a right to expect from the soldier is, undaunted bravery in the first attack; but obstacles unsurmountable are not to be thrown in his way.—After a repulse, perseverance depends on circumstances; where there is not a great probability of prevailing, the multitude will not greatly exert themselves: the few instigated by principles of honour are capable of it, but the many must have their prospect of success pretty apparent; every method must be taken to ensure it, by order, discipline, and example.

*Soldier.*

---

*Movements of March.*
The movements of march used in going from one camp or quarter to another, are made in one or more columns; and are generally slow, as being combined with those of cavalry, artillery, baggage, and with the nature of the country. In this situation such additional precautions are taken, as to be in no danger, from attempts of the enemy, who are then commonly at a distance.

*Movements of manœuvre.*
The movements of manœuvre are such as are executed near, or in presence of an enemy; and have for their object the formation of the line, in order to make the attack in the most advantageous manner that circumstances will allow of—Therefore celerity in change of place cannot be too great, when consistent with order, as it gives the enemy less time to see the intention of, or to counteract such movements.

---

The movements of all great bodies are made either in line or column; and in both cases are regulated by some one given division, as that of general direction.

In column, the leading division is always that of direction, and is conducted by the commander himself.

In line, some one battalion, according to circumstances, and the views of the commander, and in general a flank one, is named as that of direction. Regulating division.

To the movement of this body the commander himself will attend, and see that its points of march, and pace, are such as he wishes to conduct the line by, and which in every respect conforms to its motions.—He will change his place, whenever he finds it necessary to name a new division or battalion of direction, and will by a caution apprize the line of such alteration.

---

It can never be too often repeated or recollected, that one or other of the flank battalions of the line, directs the operations of the whole. To its movements every other part yields, and conforms; there are very few cases in which the center ought to regulate. In line, the flank battalion regulates its movements.

If an enemy is to be turned, or an attack made, it is by the flank that such movements are led. It is the flank that must preserve the point of *appui* in all movements in front. If the line is thrown into column, it is the head, or leading flank of that column which conducts, and whose writhes and turnings are followed by every other part of the body; and such head becomes a flank when formed into line.—It is seldom that an attack is formed from the center, but a movement never need, nor should be, although (relative circumstances not considered) it is certain, that in the march of the line in front, it appears to be easiest conducted by the center.

The commander will therefore be with whichever flank directs the operations of the line, and by which he proposes to make the attack, or counteract the attempts of the enemy.

---

All marches, and movements must be made in given lines, and all formations on fixed and determined objects; and those lines and objects must be correctly given before the movements or formations are begun.—Intermediate points, where heads of columns are to arrive, or where flanks of lines are to be placed, must be carefully ascertained by officers advanced on purpose.

Commanding officers of a line must always be assisted by several intelligent and detached officers, and non-commissioned officers, who Movements made on fixed points.

should be employed occasionally in ascertaining points and distances. The method of finding intermediate points, betwixt two given distant and inaccessible objects, must be practised by every officer, and often employed in formations of the line.

<small>Mounted officers.</small>

The superior staff officers of the battalion, must be at all times mounted; to give ground in movements; speedily to correct mistakes; to circulate orders; and above all, exactly to dress the pivots of divisions on the distant forming points, when the columns halt and are adjusted, before wheeling up into line; an operation, which no dismounted officer can effectually perform, in an extended line.

---

The exercise of small bodies, when within the command of one voice, appears more showy, by keeping such bodies constantly in motion, and by changing from one manœuvre to another while on the march. But such movements, and the formations made from them, must be on accidental points, and however brilliant in regimental practice and review appearance, where the lesson of the day has been previously arranged, they can only be considered as occasional exceptions, not applicable to large bodies, where hurry must be avoided, and where concert and relative position are indispensable.

<small>All changes of position originate from the halt.</small>

If the principle of moving, forming, and dressing, upon given and determined points is just; all quick alterations of the line while upon the move, and not proceeding from a previous halt (however short) are false and defective: the effects of which, though not so apparent in a single battalion, would be very obvious in a line or column of any extent.——A pause between each change of situation, so essentially necessary in the movements of great bodies, should seldom be omitted in those of small ones: squareness of dressing, the exact perpendiculars of march, and the correct relative movements of the whole are thereby ascertained.

All alterations of position from one situation to another, made by considerable bodies, must begin from a halt, except giving a new direction to the heads of columns, or diminishing or increasing their front, which may be done while in movement.

Alterations

Alterations of position made from the halt, may succeed each other quickly. No unnecessary time need be taken up in scrupulous dressing, but every one be instantly apprized of the following movement. *Alterations of position made from the halt.*

---

Words of command, for all movements of lines or columns, must be given— *Words of command.*

In such manner, and in such extent of voice, that every individual should be apprized of what is to be done. *Explanatory*

Short—Quick—Loud—so as to be caught and repeated from right to left of a line, or from front to rear of a column, in the shortest time possible. *Executory.*

All alterations in carrying arms, change of pace, facing, inclining, halting, marching, and in general every operation of the battalion, whether in line or column, which ought to be executed by the whole, at the same instant, are made in consequence of one word from the commanding officer. But when broke and in column on many occasions, the leaders of divisions repeat or give the words of—march—wheel—halt, &c. to their several divisions as is necessary.

Every individual is apprized by a caution of what is intended to be done, before any alteration of position takes place. *Commands given.*

The commandant of the line or column, gives the first explanatory direction to the battalion or division, which then regulates, and from thence it circulates.

Every movement must be divided into its distinct parts, and each part executed by its explanatory and separate words of command. *Division of movements.*

No time is lost in complying with these directions; confusion, hurry, and uncertainty will be thereby avoided; precision, order and uniformity attained.

On all occasions when words of command are not heard; if the directing division has made a change of situation, the rest of the body will conform to it, as soon as the intelligence of the officer, has pointed out what is meant to be done. *Division of commands.*

Every officer muft practife giving his words of command, even to the fmalleft bodies, in the full extent of his voice, and in a fharp tone. By fuch bodies he muft not only be heard, but alfo by the leaders of others, who are dependant on his motions. The juftnefs of execution, and the confidence of the foldier, can only be in proportion to the firm, decided, and proper manner, in which every officer of every rank gives his orders.

*Neceffary attentions.*

In the midft of furrounding noifes, the eye, and the ear of the foldier, fhould be attentive only to his own immediate officer; the loudnefs of whofe commands, inftead of creating confufion and unfteadinefs, reconcile to the hurry of action, which is generally fo oppofite to the ftillnefs of common exercife. All fuperfluous or drilling directions are improper, when the regiment or line is together; and every one fhould be alone attentive to the exact execution of his allotted part.

The eye will often point out the propriety and moment of conforming to the movements of the regulating body, when the ear does not receive the explanatory word of command.

*Signals.*

Signals are improper in exercife, becaufe dangerous and apt to be miftaken in fervice: every direction ought to proceed from the voice, which is explanatory to the underftanding. But a few well diftinguifhed fignals of the drum or trumpet, may on fome occafions be permitted as expreffive of the alterations of pace, and as preparations to a march—halt—quick ftep—flow ftep—forming line from column. But thefe which come only from the commanding officer, are to be confidered as addreffed to the leaders of battalions or other bodies, who are in confequence to give their verbal orders for the execution.

On all occafions of parade, the drums and mufic add much to the appearance of the troops; but they are improper in manœuvre, and counteract the regular movements of great bodies. They are conftantly varying the times of march, they create noife, prevent that equal ftep which habit alone can give to troops, and tend to

deftroy

destroy the very end they are meant to promote; for the uncertain time of an inftrument, can never regulate the cadenced march; and the fame found is progreffively heard at different inftants, in proportion to the diftance of the point from which it iffues; as this perceptibly takes place even in the extent of a battalion, its operation on a greater body or on a line, is evident. The tact or cadenced march, can be preferved from the eye and habit alone, and troops muft be accuftomed to maintain it, notwithftanding drums, mufic, or other circumftances may be offering a different marked time. Their general ufe therefore is on occafions of fhow and parade; at the moment of the charge, they may however be allowed, as infpiriting and directing the attack, and alfo in the column of march; but in moft movements of manœuvre, during every tranfition from line to column, or from column to line, in all formations, and in the march of the line in front, they muft be fparingly ufed, and never as directing the cadence of the ftep, or in the inftruction of the recruit, officer, or battalion. *(Mufic, drums, &c.)*

## Points of March and Formation.

Every leader of a body, which is to move directly in front, muft take care to conduct it in a line, truly perpendicular to that front, otherwife, it can never preferve its order intire, but muft clofe on one flank, and open out on the other.

To march on one object with certainty, and without wavering, is exceeding difficult, and not to be depended on—two objects therefore, placed in the fame direct line, are neceffary for the purpofe of correct movement. *(Points of march.)*

This rule of marching on two objects, is general, and muft be obferved by the leaders of all bodies, great or fmall, whether in front or file.

Therefore, before fuch movement is begun, a principal diftant and remarkable object, perpendicular to the front, and to the point by which the body is to move, muft be chofen. *(Fig. 1.)*

The

( 30 )

The leader will then fix on some intermediate stone, tuft of grass, or whatever occurs on the surface of the ground (and such always occurs) and will conduct his body, so as to preserve these two objects in the same line.

From time to time as he approaches the old ones, he will choose new intermediate points; and by this means will be able to advance in the same direct line with a mathematical exactness, which is essential in the movements of great bodies.

When necessary to alter the direction of any body, new points of march must be ascertained; and by wheeling or other operation, such body must be placed square with such points, before it can properly move upon them.

*Points of march.*

*Fig. 1.*

A person placed in the front of any body, can with difficulty determine the line which is perpendicular to such front—but, if he posts himself in the rear of it, he can then, having the square of such body before him to direct his eye, more easily ascertain the perpendicular. It is therefore from the rear that such points should be given; and could we suppose ranks and files most perfectly correct, the prolongation of each file, would be a perpendicular to the front of the body.

---

*Points of formation.*

As the points of march, are those on which bodies move, and are always perpendicular to the front; so the points of formation are those on which they are at any time formed, and are on the flanks in the prolongation of the front, after such formation.

The more remarkable and detached those points are from other objects, the better.

## Of the POINT D'APPUI, or POINT of SUPPORT.

The point of APPUI is that to which the wing of a battalion, or of a more considerable corps is placed. From that fixed point when the body forms, must it begin to line itself on the opposite point

point of formation. Therefore, the *alignement* is taken from the right wing, when broke from the right; and from the left wing, when broke from the left; becaufe the body is fuppofed marching to the point of *appui*, to place itfelf there. One cafe is an exception from this principle, when diftances are taken from the rear of the column.

Fig. 1.

As when moving, or when halted, one of the flanks of every body follows out a prefcribed line, or refts on a given point; dreffing is therefore always to that hand, as to the line or point of *appui*. And this great principle is to be obferved from the fmalleft body to the moft confiderable corps, and regulates the combined attack of the army as well as the march of the platoon.

Point of *appui*.

The exact dreffing of the battalion is one of the moft important of evolutions; and the proper advancing or retiring of the line depends upon it. The beft chofen and well trained foldiers will fall into diforder, if they remain ill dreffed in their battalion.

---

It may be fuppofed, that the faults which fpring from a defective *alignement*, can be corrected while in motion—this may be done with a fingle battalion, but is nearly impoffible with a line compofed of feveral. It will be always beft to drefs it, as it ought, before marching: and every officer muft therein affift, as the moft effential requifite to advance well in line.

For this purpofe it is neceffary—

That officers always drefs the faces of the foldiers of their divifions, and not their breafts.

That an officer fhould never drefs a body without determining, or at leaft fuppofing a line on which the divifion or battalion is to be formed.

That he ought conftantly to ftop at the point chofen for his point of *appui*, and to which his flank is to be placed: From thence he will drefs till he has dreffed his platoon, or at leaft a part, on the diftant point of formation. If he does then move, he muft always confider the laft man whom he has fo correctly dreffed as the point of *appui*, and proceed accordingly—But when employed in the act of dreffing,

Dreffing from the point of *appui*.

dressing, he must not run along the division; for while moving, he cannot preserve his intended direction, and a false *alignement* will thence arise.

In the same manner is a whole battalion dressed; the divisions of which may be considered as the individuals were in the division—The flank of the first is not to be quitted, till the second division, or at least part of it, is dressed; nor must the flank of the second be quitted, till the third is dressed.

## The ALIGNEMENT.

THE ALIGNEMENT is the straight line drawn between two given points, either of march or formation. To march, or form in the *alignement*, is to make troops either march or form in that straight line.

*Fig. 3.*

On the justness and observance of this line depends the accuracy of all movements and formations, and therefore every relative help must be applied to ensure it.

*Points fixed in the alignement.*

1st. The general direction of this line is always determined before the troops arrive on it.—2d. *The point of that line at which the head of the troops is to arrive, must next be ascertained.*—3d. That line must be accurately traced out and subdivided, in order that the troops may march and form correctly upon it.

Therefore before a column open or solid, approaches the ground where it is to form—the commander will ascertain, as circumstances may determine him, the advanced and distant points, at which the flanks of his line are to be placed, or which he intends to be in the prolongation of his line when formed.

If he enters his *alignement* at one of those determined points, he continues his march straight upon the adjutants and the other.

But if he enters the *alignement* somewhere betwixt them; it then becomes necessary to ascertain the point where the direction of the march intersects the new *alignement*, and which is the point where the head of the column first arrives in it.

The following method of finding an intermediate point between two given, and perhaps inacceffible objects, muft be thoroughly underftood and conftantly practifed: and if one column in this manner afcertains its point, any other number connected with it in their movements, will in fame manner, or eafier determine theirs.

Fig. 4.

When the head of the advancing column approaches whatever part of the ground it ought to arrive upon—Two officers A, B are fhown the flank points of the *alignement*, D P, and are fent forward to determine the intermediate point O, at which the head of the column ought exactly to enter into the new direction.

Method of finding intermediate points in the *alignement*.

They feparate feventy or eighty yards, go to the fide to which the column is to wheel, and B immediately places himfelf in the line of A P, advanced before the head of the column—They then both move on, B always preferving A in a line with P, and each defcribing the portion of a circle upon P, as a center—A looks to B, and moves on while the point D continues to be advanced before him: but the inftant he has brought B in a line with D, they both halt; and the four points are then in the fame line.

Fig. 4.

A remains fixed—B returns to the column, which has been purfuing its march in its given direction; and when it approaches nearer, he having now the line of march and the *alignement* both afcertained to him, again advances, and eafily places himfelf at their interfection O, which is the exact point where the head of the column is to arrive.

The officers employed to give the *alignement*, may in this manner exercife themfelves, and determine it at the gallop, and therefore no halt or ftop of the column is to be apprehended; and all officers whatever, ought to be perfect mafters of this operation, as ufeful in the conduct of the moft minute divifions, and particularly in preferving the flanks of battalions when in line, dreffed in the general direction of the feveral colours.

When two bodies are in march to gain the fame given point; the above method may moft ufefully be applied, to afcertain which of them can firft arrive at it.

**Method of determining which of two columns arrive first at a given point.**

Fig. 4.

The column C, and enemy E, are both in march on the point O. The leader of C, obferves a diftant point at S beyond, and a-head of the enemy E. If he can continue to keep this object open, and in front of the enemy, it is a certain fign that he approaches fafteft to his wifhed for point; but, if it appears moving towards the rear of the enemy's march, it indicates his advantage, and the attempt muft be given up in time.

---

When a number of fixed and confpicuous objects happen to be found in the *alignement* in which the column marches, no farther helps feem neceffary for preferving the pivot flanks of the divifions compofing it, in that line.

**Points in the *alignement* beft afcertained by adjutants and mounted officers.**

But, as there are many fituations where heights and valleys intervene, and where no remarkable objects, diftant or intermediate, occur in the direction, which (perhaps on account of that of the enemy) muft be chofen; the moft general and fureft points to move on, on all occafions, are the adjutants, when trained for that purpofe.

As the advanced adjutants, are always front points when marching in the *alignement*; fo the point of wheeling or entry into the *alignement*, is always a fixed rear point (when more marked ones do not offer) on which to regulate and correct the march of the column, after it has entered on the *alignement*.

Fig. 3.

Suppofe O P, are two given points in the *alignement*—at O, the line is entered upon; the firft adjutant of the general or of the leading battalion advances towards P, not more than one hundred and fifty or two hundred paces, otherwife he might retard the march of the column. He is lined on the point P, by the commander or by fome one placed at O; he then marks the point over which his body is, and retires his horfe, till its head is precifely over that point—the adjutants turn their faces towards the column when it has entered the line.

In this manner, one intermediate point will have been found, viz. the head of the horfe, which the divifions touch in marching, and which they preferve in a direct line with the point P. But when the officer of the firſt diviſion has arrived at that adjutant, he will find himſelf in the ſame ſituation as when he was at O; conſequently, another intermediate point muſt by this time be prepared for him.

The adjutant of the ſecond battalion of the line, or another aid of the general's, advances in ſufficient time, and remains with his horſe's head where that of the firſt adjutant was—he immediately moves on about two hundred paces, lines himſelf with the ſecond adjutant and the point O, then places his horſe's head as before directed; and thus a new intermediate point is given. If the line extends farther, a third adjutant relieves the ſecond, he the firſt, who anew arranges himſelf on the horſes heads of the other two, and proceeds as already directed.

*Points in the alignement how aſcertained by adjutants and mounted officers.*

Fig. 3.

Three adjutants will be ſufficient to give the *alignement* to a whole line, and to keep them in it, if they thus alternately relieve each other: but if the line is very extenſive, it will be proper to detach one or two more to aſſiſt and relieve them, in order to prevent the center and rear from loſing the *alignement*.

In this manner, and in the thickeſt weather, when neither diſtant points, nor the extent of the body can be ſeen, do the moſt conſiderable columns of Pruſſian troops preſerve, and form in a given and determined direction; and in this manner is a line of march or formation prolonged to any length, and ſuch direction given it, with or without the help of fixed points, as beſt ſuits the circumſtances of the movement, and the views of the commander. Another advantage alſo ariſing, is, that though the diſtant point may not be ſeen or known by the pivot leaders, yet the mounted officers always ſtanding faced to the new line, and with their horſes heads directly over it, are known by every one to be there placed for the purpoſe of marching or forming upon— three ſuch officers ſhould be always ſtationary, at the ſame time, in a line of any conſiderable extent.

---

Although when two diſtant points are given, a perſon placed at one of them, can direct an intermediate point in the line of the

**Method of *alignement* from one to another given point.**

other: yet he will do it with more exactness, if instead of remaining at the first point, he retires a few paces from it, and looks along both; in this situation, he will have his eye more correctly guided in the line of the two given points; as it is in fact obtaining another object of direction. This circumstance must be attended to, when the line is to be prolonged, intermediate points found, or in the dressing of all bodies where the flank point, and one other distant point is given. If there are several distant points given in the same line, this precaution does not become necessary.

---

The INSTRUCTION of the officer, the recruit, and the squad, so as to qualify them to discharge their duties in the greater bodies of the company and battalion, comprehend a variety of important and necessary articles, each of which must be particularly detailed. Their GENERAL HEADS are as follows;

## OFFICER.

EACH officer must be taught every individual circumstance necessary to a recruit, accustomed to give words of command, and the exercise of the fusil, sword, and esponton.

**Instruction of officers.**

Squads of officers, should be occasionally marched and exercised by a field officer; they should be much practised to judge of and preserve distances, both in front and file: to cover pivot flanks; to take up and prolong lines; from the number of files in their divisions, to be accurate in the ground necessary for each, and to extend that knowledge to the front of greater bodies; to ascertain perpendiculars and the squareness of the wheel; to be decided, pointed, and alert in their commands and movements: habit and attention

attention will foon produce perfection, in thefe and every other requifite; and individuals from time to time when qualified, muft be ordered to exercife the battalion, or its parts.

The complete inftruction of an officer, enlarges with his fituation, and at laft takes in the whole circle of military knowledge. From the variety required of him, his exertion muft be unremitting, each to make himfelf mafter of his own part. We have already fhown the neceffity of this, and that he is the fpring, which regulates or diforders the great machine.

## R E C R U I T.

RECRUITS muft be individually taught in the following progreffion.

Without Arms.
- Pofition.
- Turning the eyes to the — — — — { Right. Front. Left.
- Facings when halted to — — — — { Right. Left. Rear.
- March on given points to the — — { Front, Flank } Ordinary Time. Oblique } Quick Time.
- Turning on the march to — — — { Right. Left. Rear.
- Halting.
- Dreffing on given points to — — — { Right. Left.

*Inftruction of the recruit.*

---

When prepared for it, feveral recruits are then joined together in a fingle rank, but at open files, that they may carry themfelves properly, be independent, and learn not to lean upon one another, or to totter, and feparate their arms from their bodies. In this

fituation

situation they march for some days, and perform together what they have been taught singly.

The accoutrements of the recruit are then put on: the firelock is given; and each man is separately instructed in the use of it.

With arms, he practises the same movements that he hath been taught without arms; and is then gradually brought forward in the several parts of the firelock exercise, both manual, and platoon.

---

When thus far individually advanced, the recruits are then assembled in squads, more or less numerous, and proceed first in single ranks, and then in bodies, to be instructed in,

*Instruction of the recruit.*

With Arms.
- The distance of ranks and files.
- Opening and closing of ranks and files.
- Dressing ⎫
- Marching ⎬ on given points.
- Halting ⎭
- Facing to the flank, or rear.
- Marching in ⎫
- Wheeling in ⎬ File.
- Fronting from ⎭
- Wheeling the quarter circle from the { March. / Halt.
- After the wheel — — — — — { continuing the march, or halting.
- Manual ⎫ Exercise.
- Platoon ⎭
- Firings { by single men of each file. / by single files. / by single ranks. / by platoons.

---

All these, and other heads of instruction for the recruit, must be followed out according to more detailed regulation: and it requires in the officers of each company, who have such care, the most constant perseverance and knowledge.

They must allow for the capacity of their recruits; be patient, and not rigorous, where good will and attention are not wanting. Quickness is not to be required at first; it is the result of much practice. The recruit must be carried on progressively—he should comprehend one thing before he advances to another. *Mode of instruction.*

The difficult motions must be practised oftner than the easy ones, and an equal command of both thereby attained. The prescribed modes must be minutely followed. In learning each motion and position, the firelock, fingers, elbows, &c. must be justly placed by the instructor; but when recruits are more advanced, they must not be touched, but taught to correct themselves when so admonished. They should not be kept too long to any particular part of exercise, so as to fatigue, or make them uneasy; but marching without arms should be much intermixed with the firelock instruction.

---

It is not meant here to enter into a minute detail of the preceding articles; but the opposite points being more immediately connected with field movement, which is the great object of this work, are looked upon as fundamental and indispensable, and unless they are observed, the whole edifice falls to the ground.

{
Position.
March { in front. oblique. to the flank. in file. }
Halting.
Dressing.
Wheeling.
Distance of { Files. Ranks. }
}

## POSITION.

The equal squareness of the shoulders and body to the front, is the great principle of the position and movement of the soldier, and cannot be too much attended to. That position should never vary, whether he is halted, or in march; and the situation in which he *Position.*

can,

can, and ought to move, fhould therefore determine that in which he is to ftand ftill.

**Without arms.** The heels are nearly clofe, the toes a little turned out—The arms hang near the body, but not ftiff—The flat of the hand touches the thigh—The elbows and fhoulders are kept back, and the belly is drawn in, which occafions the breaft to be advanced, and the body to be upright, but rather inclining forward—The head fhould be erect, and not turned (or the leaft poffible) to right or left, as the fhoulders would certainly accompany it; but the eyes only being occafionally glanced, will enfure correct dreffing, if the body remains fquare.

---

**With arms.** When the firelock is given, he is firft taught to carry it fhouldered. In this fituation, his body remains in the fame exact pofition as before prefcribed, except that the wrift of the left hand is turned out the better to embrace the butt, and the left elbow a little bent, without being feparated from the body. The firelock is placed in the hand; and carried in fuch a manner, that it fhall not advance or keep back one fhoulder more than the other; the butt muft therefore be forward, and as low as can be permitted without conftraint; the barrel part muft reft in the hollow of the fhoulder; the left elbow muft not be drawn back, or the firelock attempted to be carried high; otherwife the one fhoulder will be advanced, the other kept back, and the upper part of the body diftorted, and not placed fquare with refpect to the limbs.

## Of the MARCH.

**The march in front.** ALL military movements fhould be made with the greateft quicknefs that is confiftent with order: the uniformity of pofition, cadence and length of ftep produce that equality and freedom of march on which depend their effect and precifion. To this, the foldier muft be carefully trained; nor muft he be fuffered to join the battalion till he is thoroughly perfected in this moft effential duty.

In marching, each soldier must be well balanced on his limbs—his arms must be steady, and not vibrate—his body must turn neither to right nor left, incommode his neighbour, nor stoop too forward; much less must he lean back, for in such position he can never gain ground in marching. He must never advance or keep back one of his shoulders: the left is generally thrown back, from a mistaken notion, that the firelock is thereby better carried—He must not turn his head to the hand to which he dresses; as a turning of the shoulder will undoubtedly follow: but he must be accustomed to turn his eyes only along the faces of his neighbours, in order to perceive the flank man.

He will never be taught to march well, if he is not instructed to stretch his ham, but without stiffning his knee; and to point his toe, keeping it near the ground, so that the person before him cannot see his shoe soals either when he raises or puts down his feet. Position in the march.

In marching as the ham must be stretched, so must the body be rather thrown more forward than in the position when halted: the first is effected by pointing the foot, and touching the ground with the toe rather than the heel: the second is effected when he understands, that the body is supported by the leg which is on the ground, and that no weight can rest on that which is in the air; this circumstance also tends to stretch the ham.

Care must be taken, that the elbows are at all times kept steady, but without constraint. If he opens them, he must press one of his neighbours; if he closes them, there becomes an improper distance, which must be filled up: floating on the march in either case arises, which is too dangerous, not to demand every possible remedy that can be applied.

---

Many different times of march must not be required from a soldier, or he will be imperfect in all—Two may suffice for all occasions; the ordinary, and quick march. The cadence of these, and the equality of step, must be habitually and strongly imprinted on each soldier. If at any time the march requires more or less ground to be gained by individuals, or by bodies; in the one Times of March.

case, it must be done by lengthning the step a little, and in the other, by shortning it; but the cadence of the then march must not be altered.

The general length of step must be such as will not demand exertion; as artillery in line can move with; as surfaces not perfectly uniform and smooth will allow of; as men of a small stature, in close ranks, and in great bodies, can conform to and keep their order perfect; and such also as will admit of being lengthned when circumstances require; that part of a line which is behind may move up, that the divisions of a column may keep or regain their just distances, or that a body, on no very extensive front, may step out for a short space on particular occasions.

<small>Length of step.</small>

The time of march must also be such as a great body can persevere and move in, without hurry or disorder.

The ordinary, or slow march, may be eighty in a minute; each step, thirty inches. It is the pace in parade, common marching in front, and on all occasions where greater celerity is not ordered. It is nearly at the rate of two miles and one quarter per hour. It is also the pace with which, in an easy and unconstrained manner, the columns of an army perform their marches without being restricted to perfect uniformity.

<small>Ordinary march.</small>

The quick march may be one hundred and twenty in a minute, each step 30 inches. It is the pace in all wheelings, and in all filings of divisions from line to column, or from column to line; and for quick movements in front, either in column, or occasionally in line when ordered. This step may, on urgent motives, be increased to one hundred and fifty in a minute; beyond that degree of quickness, it becomes a run.

<small>Quick march.</small>

A division or company may occasionally run, a battalion may march quick; but the hurrying of a column or large body in front, will certainly produce confusion, and disorder. It is never to be risked where an enemy is to be encountered; though it may sometimes be necessary where a post, or situation is to be seized.

Although the length of step in movements in rank to the front, is thirty inches; yet in movements in file, it will be found necessary

( 43 )

to shorten it: it must therefore on such occasion be made at twenty-four inches, in order to insure correct distances. *Steps shortened in defile.*

To these degrees of march, the soldier should be accustomed without drum or music, and by constant practice taught to acquire the given times and length. Plummets, vibrating the above numbers in a minute, should be in the possession of every instructor, and will prevent the uncertainty which universally takes place; they may also be advantageously employed in regulating the music. Accurate distances of steps should be marked out on the ground, along which the recruit should march, and acquire the just length of each. *Mode of instruction.*

Length of plummets nearly, and vibrations in a minute 
$$\begin{cases} \text{Inches} & \\ 39 \quad - \quad - \quad 60 \\ 28.5 \quad - \quad \phantom{-} 70 \\ 25 \quad - \quad - \quad 75 \\ 22 \quad - \quad - \quad 80 \\ 9.8 \quad - \quad 120 \\ 6.2 \quad - \quad 150 \end{cases}$$ A rod or cane of an uniform thickness, and one-third longer than a plummet, will vibrate in the same time as the plummet.

Marking the time of march by continuing the cadence and movement of the feet without advancing, should be taught and practised: it is useful in dressing the battalion before halting; and in waiting for the rest of the line, when any particular battalion has overstepped it. *Marking the time.*

It has been customary from the halt, when to march in front to step off with the left foot; and when marching in file, to step off with the foot next the proper front. *Beginning the march.*

---

To preserve equality of front, to prevent opening out, or closing in, and the consequent confusion that must follow; the march of every body, great or small (except in the case of inclining) is necessarily made on a line perpendicular to the then front of that body; and as its component parts should therefore move on that perpendicular, or on so many lines parallel to it; each individual must in his person be placed and remain perfectly square to the given line, otherwise he will naturally and insensibly take a direction perpendicular to his own person, and thereby open out, or close in, according *Attentions in marching.*

cording to the manner in which he is turned from the true point of march.

If the diſtortion of a ſingle man operates in this manner, what muſt that of ſeveral occaſion, each of whom is marching on a diſtinct front, and whoſe lines of direction are croſſing each other? they muſt inevitably produce confuſion!

From the want of a due attention to this circumſtance, and to the equality of the march, ariſes that floating, diſorder, and incapacity of movement in bodies, great and ſmall, which we often ſee, and which is attempted to be remedied by the looſeneſs of files, and other dangerous and improper expedients; but which can only be obviated by the accurate poſition above required.

This circumſtance, chiefly combined with the equality of the march, the touch of the files which is never relinquiſhed, juſt diſtances, and the true lines on which the Pruſſian troops move; give to them without apparent conſtraint, the head being turned, or the leaſt trouble taken in dreſſing, the moſt wonderful and deciſive accuracy in the marches and operations of the greateſt bodies.

## The March in File.

*The march in file.*

THE whole muſt move at the ſame inſtant, each replacing the feet of the man before him, and no opening out muſt on any account be permitted. The front rank will march ſtraight along the given line: each ſoldier of that rank muſt look along the necks of thoſe before him, and never to right or left; otherwiſe a waving in the march muſt take place, and of courſe the loſs and extenſion of line and diſtance, whenever the body returns to its proper front.

The center and rear ranks muſt look to, and regulate themſelves by their leaders of the front rank, and always dreſs in their file. In general, the ſoldier muſt be accuſtomed to obſerve and keep his file; and muſt underſtand, that he is placed in that manner for the ſole purpoſe of ſuſtaining his leader; and each three men, ſhould conſider themſelves as a complete body, ſo arranged for the purpoſe of effectual reſiſtance.

As all file marching is in general made in quick time: the rank or division, after facing to the flank, must at the word *march*, instantly step off, observing what is before directed, but necessarily diminishing the length of step—the soldier must be well exercised to this pace, much irregularity and time spent in adjusting distances, will be thereby avoided.

## The Oblique, or Inclined March.

To this essential movement, necessary on most occasions, the soldier must be dressed with the utmost care. He must understand that as in marching directly forward, his person is perfectly square with the line on which he marches; so in inclining, he does not alter that square position of his body, but is carried in the oblique line, by the mere movement and crossing of his legs. *[The oblique march.]*

When for example, he is to incline to the right in rank or division, he must carry his right foot to the right, and at the same time advance in the diagonal line, in which the body is to incline: the left heel must then be placed before the right toe—this operation is repeated till the incline ceases, and the division is ordered to resume its former parallel front. *[Attentions in inclining.]*

But in the course of this operation, the soldier is naturally induced to advance his left shoulder, and keep back his right; a fault which must be absolutely avoided, as inevitably altering the direction of the front, and tending to confusion in a large body. The officer must therefore take care, that the other extreme is not fallen into, when he directs the right shoulder to be kept forward, and that each soldier understands this order as a direction only, not to let it be kept back, but that his whole body and feet should remain in the same position they were in before he began to incline. The officer himself must always recollect, that at each pace which his division makes, the whole of it remains on a line parallel to that from which it set out.

Divisions will be accustomed to turn their eyes both to the hand to which they incline, and also from it; as it is often necessary ceffary in line for a battalion to incline, the whole continuing dreffed to its center.

**Degree of obliquity.** One certain degree of obliquity muſt be underſtood, at which the body is always required to incline; this will naturally be in the diagonal line between the front, and the perpendicular raiſed upon it; ſhould other directions of march be occaſionally required, it ſeems then difficult to anſwer for the preſervation of the front, which is an indiſpenſable article.

---

**Inclining in file.** Diviſions muſt be accuſtomed when in file to incline to either flank, in the ſame manner as is directed when in front, each file looking upon itſelf as a line which inclines in front; and as this in many ſituations, particularly in flank marchings is highly uſeful, an officer cannot too much train his diviſion to it.

## Of the SIDE MARCH.

The ſide ſtep is very neceſſary on many occaſions, when halted; for diviſions to open or cloſe to either flank without facing, or for one or more battalions to move a ſmall diſtance to right or left, in line, and without changing their front.

At the word *march* (if to cloſe to the right) each ſoldier at the ſame inſtant, puts his right foot directly to the right ſide in the given line, and draws his left foot after him, till the whole **The ſide or cloſing march.** arrive at the intended point—in doing this it muſt be obſerved,

That the flank leader takes ſmall but quick paces.

That the ſoldiers as in inclining, preſerve their ſhoulders equally and fully to the front.

That they preſerve their relative diſtances.

That in this movement they do not bend their hams.

That they do not paſs the given line.

That the diviſion looks to the hand it cloſes to.

That no ſoldier on this occaſion, if he has loſt his diſtance, is to halt till he has recovered it, though his officer may have given the word *halt*.

## Of the HALT.

As at the word *march*, the whole body step off together; so at the word *halt*, when marching in front or file, the whole must halt at the same instant. The foot then in the air, finishes its proper pace; the other foot is brought up to it, and the whole remain looking to the point by which they were marching, till otherwise ordered by an after command.

All halts are made to the point to which the troops are then looking, and a separate command directs the after dressing.

## Of DRESSING.

At the word *dress*, each individual remains in his true position, the eyes are cast towards the point to which they are ordered to dress, and the smallest turn possible of the head allowed, in order to facilitate it; the whole person of the man must move as is necessary, and bending forward or backward, not be allowed. The faces of the soldiers are the line of dressing, not their breasts or feet; a man who has a broad or raised chest will otherwise be behind his neighbour, who has a thinner one.

No rank or body ought ever to be dressed without the officer on its flank, determining or at least supposing a line, on which the platoon or battalion is to be formed.

In the instruction of recruits, all marching, halting, and dressing, must be made on given objects (at least two) in the same direct line; and accident must never be allowed to direct such operations.

When the word *dress*, is alone given, it means to the hand to which the troops are then looking. When eyes are to be turned to another point, it will be expressed by the addition of *right*, *left*, or *center*.

## WHEELING.

**Of wheeling.**

To wheel, is to give to a division the same position, which a single soldier takes, when he faces towards either flank. The man does it by turning his body, the division by describing the ordered portion of a circle.

A single rank or division may wheel at any pace prescribed, and without a stop or alteration of the time, at which it is then moving. But when a number follow each other in column, and are to wheel above one-sixth of the circle successively, on the same identical ground; it becomes necessary, in order to prevent false distances, and a lengthning out of the column, that the divisions make their wheels at a pace considerably quicker, than what the body of the column is then moving at. Therefore, when marching in ordinary time, the wheels must be made in quick time; and if marching in quick time, they must be made at a still quicker rate.

**Mode of wheeling.**

*Uniformity of front, is to be maintained during the wheel,* therefore the whole look to the wheeling flank; the same time is observed by each man, but his step is shorter in proportion, as he is near the standing flank, on which the wheel is made. The whole remain closed to the standing flank; that is, they touch without incommoding their neighbour; they must not stoop forward, but remain upright, and the hams are extended as when marching in front—the outward wheeling man looks to his rank, which serves him as a radius, with which as accurately as he can, he describes the required portion of the circle. Opening out from the standing flank, is to be avoided; closing in upon it during the wheel, is to be resisted.

When the wheel is compleated, the general march at which the body was moving is resumed, and dressing is ordered by the proper point in advancing, or halted.

The necessity of successive divisions wheeling in an increased time, being apparent; whenever therefore, the leading one arrives at the point where the wheel is to be made, it receives the word *halt*; then instantaneously *wheel* (to whatever hand it is to be directed); *march*; on which the eyes are turned, the whole move, and the wheel is compleated—*halt*, is then given—*eyes* to the pivot flank (if they are not then looking to it) and whenever the second division, which has continued to advance in ordinary time, arrives close on the wheeling point, the word *march* is given to the leading division, and it moves on so as not to occasion a momentary stop to the division behind it, which at that instant, and in the same manner, receives its words—*halt—wheel—march—halt—eyes*, &c. and *march*, whenever the leading division has gained its proper distance from it.

In this manner they succeed each other; and if the words of command are justly given, the wheels performed at an increased time and step, and the proper pause made after the wheel, no extension of the column will take place, but the just distances betwixt divisions will still be preserved.

The rear ranks are always closed up, and during the wheel, they incline and cover their proper front rank men. All the divisions must take care that they march correctly upon, and wheel exactly at the point where the leading one wheeled, and do not shift to either flank, which without much attention they are apt to do.

It often happens that soldiers of the division, and particularly of the flanks, halt before the officer has commanded *halt*, or overwheel some paces, although he has already given his orders: this happens from the soldier being so much accustomed to wheel the quarter circle, that he mechanically stops whenever that is compleated. It is therefore difficult for an officer to give his division a direction, under or above that portion of the circle; and it is of consequence, that they should be much trained to unequal wheel-

*Attentions in wheeling.*

wheelings, the beginning and ending of them, being only determined by the command of the officer.

---

<small>Wheels muſt be directed by the moving flank.</small>

Wheeling by looking to the ſtanding flank, is a modern method, firſt uſed by ſome of the light cavalry, and adopted among the other looſe practices of our infantry, as arriving by a run quicker at the given point, and not ſhifting the pivot. But it is rather a formation in file, than a wheel, will ſeldom be made exact, the uniform front of the body is broken, which ought to be held ſacred, and it is not at every inſtant under the command of the officer, who ſhould have it in his power to halt it on ſuch degree of wheel, as he finds proper. Cavalry, from the eaſe with which each horſe alters his pace, may preſerve uniformity of front while wheeling in this manner, but infantry cannot. This practice was introduced, when open files took place in our battalions; becauſe the touch of the files, the principal guidance being then gone, it became difficult to preſerve what was before eaſy, the uniformity of front, during the wheel.

---

<small>Wheeling by files.</small>

The ſquad or diviſion when marching in file, muſt be accuſtomed to wheel to either flank, each file ſucceſſively, without loſing or increaſing diſtance. On this occaſion, each file makes its ſeparate wheel, but without an alteration of time of march, or the eyes of the rear ranks being turned from their front rank. The front rank men (whether they are pivot men or not) muſt keep up to their diſtance, and the wheeling man muſt take a very extended ſtep, and loſe no time in moving on.

---

<small>Formation on two ranks and at open files.</small>

The METHOD almoſt univerſally adopted in our infantry, and in ours only, of forming two deep, and at open files, deſerves the moſt ſerious conſideration. It was not produced by the experience of the German war, but by that of the firſt American. The deſultory ſervice there carried on by ſmall bodies of men, and the then deficiency of movement, and want of flexibility in our ſolid battalions,

battalions, made us run into the other extreme, and firſt introduced it as proper for that country; review appearance continued it; and the new military modes, brought into faſhion by the light infantry, have tended to make it the prevalent order of the ſervice. Many reſpectable officers are ſatisfied of its propriety; but it ſeems neceſſary to conſider its operations and conſequences, when extended to larger bodies than the ſingle battalion.

Its advantages are ſaid to be,

**That** as infantry ſeldom or never ſhock with bayonets, all formations on a great depth are unneceſſary.

That as fire now decides; the more men that are thus uſefully employed, the better. <small>Reaſons for a thin order of formation.</small>

That the fire of a third rank is thrown away, and more incommodes the front rank, than it does the enemy.

That at cloſe files, men have not the uſe of their arms; and are apt to crowd, double, and get into confuſion, when under the enemy's fire.

But it muſt be recollected,

That if theſe reaſons are good with reſpect to two ranks, ſome of them operate in favour of a ſingle rank. That though infantry do ſeldom mix with bayonets, yet it would more frequently happen if two ranks were oppoſed to three, and the conſequence ought not then to be doubtful. That rank firing, or reſerving the fire of the third rank, obviates the inconvenience complained of. <small>Reaſons againſt a looſe order of formation.</small>

That no general could manage, or poſition contain a conſiderable army formed in this manner; even one of twenty thouſand men, would occupy five miles in front. That the great ſcience and object of movement being to act with ſuperiority on certain points; it is never the intention of an able commander to have all his men at the ſame time in action; he means by ſkill and manœuvre to attack a partial part, and to bring the many to act againſt the few; this cannot be accompliſhed at open files and two deep.

That the experience of other ſervices, does not ſhow the inconveniences we complain of, ſuch as to induce them to adopt our methods, and to give up the fire of the third rank. That the third rank serves to fill up the vacancies made in the others in action; without it the battalion would soon be in a single rank. That most of our calculations seem to be for the attack and pursuit of a timid enemy, but not for defence if vigorously assailed.

That a first line thus formed, would undoubtedly give way; and if once pierced, there are few instances of an action re-established by the efforts of a second line. That it is better to have one substantial and compact line to prevent the mischief, than several redoubled and thin ones to endeavour to repair it. That no such order could in any shape oppose the attack of a determined cavalry.

That in all other services, they adhere to the old mode of files touching; and each soldier is impressed with a religious observance of never relinquishing the touch of his neighbour: by this the idea of the necessity of order, mutual support and effort, is strongly felt and observed. The inconvenience of very close files chiefly operates when the line is marching in front, which will never take place for any great distance, as all considerable changes of position are made on a reduced front, by division marchings or filings, in column or echellon.

That it cannot be doubted, when a battalion is arrived at its object of attack, at close files; that both its impulse and quantity of fire, in the same extent of front, is greater, than when the files are more open; and at any time, it is more eligible to have a division obliged to fall out of the line and double, than to have openings in it, where the enemy must certainly penetrate.

*Consequences of open files.*

The perfect and correct march of a battalion or line formed at open files, seems impossible; because its principal guidance, the touch of the files, is gone. Each man is necessarily employed to preserve a required distance from his neighbour; he is obliged to turn his head for that purpose, this distorts his body, a constant opening and closing takes place, the whole move loose and unconnected. If this must necessarily happen in the regulating battalion, its influence on a line may be easily imagined, and also the condition in which it will arrive near an enemy; who, if he is formed

at

at close files, if his dressing and line are chiefly determined by the touch, if the eyes alone are glanced towards the center, if the figure of each individual is full to the front, if the whole move square along their just lines without crowding, at an uniform and cadenced step which habit has given, will, at every instant of movement or attack, be firm, united, and animated with that sense of his own superiority, which perfect order, and due consistence will always give.

On the whole therefore, the old ideas of firmness, compactness, and mutual support, should be restored and held sacred; the formation in three ranks and at close files but without crowding, should be adhered to, as the fundamental *order*, on which the battalion should at all times form and march; and the other, in two ranks and at open files, should be regarded only as an occasional exception that may be made from it, where an extended and covered front is to be occupied, or where an irregular enemy, who deals only in fire, is to be opposed.

<small>The formation on three ranks and at close files to be preferred.</small>

---

Each soldier, when in his true position under arms and in rank, must just feel with his elbow, the touch of his neighbour to whom he dresses; nor in any situation of movement in front, must he ever relinquish such touch, which becomes in action the principal direction for the preservation of his order.

<small>Distance of files.</small>

---

There are two distances of ranks, open and close; when open they are three paces asunder; when close, they are one pace; and when the body is halted and to fire, they are still closer locked up.

Close ranks is the constant and habitual order, at which troops are at all times formed and move. Open ranks, is only an occasional exception, made in situations of parade.

The distances of files and ranks relate to the trained soldier; but in the course of his tuition, he must be much exercised at open files and ranks to acquire perfect independence, and the command of his limbs and body.

<small>Distances of ranks.</small>

The

*Firelock and firings.*

The particular use of the firelock in all its branches, and the nature of the firings that should be ordered and practised, are most important articles, and require the most minute and considered detail.

---

The foregoing instructions for recruits, are meant to enforce the necessity of uniformity in training, and the perfect preparation on just principles of each individual, for the part he is to act in the great body—till he is fully qualified, he must not be allowed to join it; one awkward man, whose person is distorted, and whose movements are imperfect, will derange his division, and of course operate on the battalion and the line in a still more consequential manner; the effect that many such would produce, is, absolute disorder and confusion.

# FORMATION

## OF THE

## COMPANY AND BATTALION.

T HE general divisions of modern European armies, are almost universally the same—into companies, battalions, regiments, brigades, divisions, wings, lines.

The just composition and arrangement of the various parts, is, what gives movement and energy to the whole.

The company and battalion are the first members.

*General divisions of armies.*

---

We can never hope to have our battalions composed on the true military principles, that direct the establishments of other nations—it were time lost to insist upon their defects. But taking them as they are, or are like to be, there is no reason why their internal arrangement and formations, should not be perfect; and such as are required in the component parts of great bodies.

*Composition of battalions.*

The fundamental order of the battalion, is, that in which it meets an enemy, the order of attack. This should be habitually impressed on the mind of every soldier, and every possible occurrence and alteration of situation must spring from it.

*Fundamental order of the battalion.*

Simplicity and uniformity, are the great principles of military arrangement. The formation of the battalion when single, must
be

**Uniformity of formation.**

be exactly the same as when acting in line—no intervals betwixt companies—all officers uniformly posted on the right of their divisions, instead of being chequered in the plausible manner of right and left—the situation of every individual efficient in line, accurately ascertained; and in cases of parade, those of staff, drummer, music, pioneer, only regarded as secondary considerations, &c. &c.

## COMPANY.

The present effective establishment of the battalion is eight companies, viz.—one light—six battalion—one grenadier, besides the staff.

**Establishment of the company and battalion.**

Each company consists of — 
$\begin{cases} \text{three officers.} \\ \text{two sergeants.} \\ \text{three corporals.} \\ \text{one drummer.} \\ \text{forty-eight private.} \end{cases}$

**Fig. 5.**

In time of war, the companies will probably be increased with one sergeant, one corporal, and eighteen private; and two additional or recruiting companies also added. But whatever number of companies are in a battalion, or their strength, or of battalions in a regiment, the following general rules must be applied.

**Ranks and files.**

The companies form three deep, and may be supposed under arms — $\begin{cases} \text{fifteen file in peace.} \\ \text{twenty file in war.} \end{cases}$

The files must touch without crowding, and will each occupy a space of about twenty-one inches.

---

**Orders of formation.**

There are two orders of formation—close order, or order of attack; and open order, or the order of parade.

Close order, is the fundamental and primary order, on which the battalion and its parts, at all times, assemble and form—open order,

order, is only regarded as an exception from it, and occasionally used in circumstances of show.

Orders of formation.

In close order—the officers are in the ranks, and the rear ranks are closed up within one pace.

In open order—the officers are advanced four paces, and the ranks are three paces distant from each other.

>At all times, the battalion or its parts, assemble and join in close order—open order may be afterwards taken for whatever inspection is necessary.

---

Each company or division forms two platoons. Each platoon forms two sub-divisions, when necessary for the purposes of march.

One officer is posted on the right of each platoon, in the front rank, covered by a sergeant in the rear rank, and no coverer in the center rank.

One corporal marks the left of the front rank of each platoon, their files are compleated in the center and rear.

Formation at close order.

The other officer, corporal, drummer, pioneer, are divided in the rear, forming a fourth rank.

When the companies are not strong under arms, and a proportion of officers are absent—in such case, officers will be posted on the right of companies only, and not of platoons; and the supernumeraries in the rear will be thereby increased.

When the line is halted, and especially during the firings when engaged—the sergeant coverers fall back into the fourth rank, and observe their platoons.

The fourth rank is at three paces distance when halted or marching in line—when marching in column, it must close up to the distance of the other ranks.

The essential use of the fourth rank, is to keep the others closed up to the front during the attack, and to prevent any break beginning in the rear; and it is a great defect in our establishment, not to allow of a larger number of non-commissioned officers being applied for this purpose.

The places of abfent officers muft be fupplied by fergeants; thofe of fergeants by corporals; and thofe of corporals by intelligent men.

Whenever the officers move out of the front rank in parade, marching in column, or otherwife—their places are always taken by the fergeant coverers, and preferved till the officers again refume them.

---

Commands { When the Company is to take open order from clofe order.

*Rear ranks take open diftance* — { At this *command*, the flank men on the right of the rear ranks, ftep brifkly back to mark the ground, on which each rank refpectively is to halt, and drefs at open diftance; every other individual remains ready to move.

*March* — — — { At this *command*, the rear ranks fall back two and four paces, each dreffing by the right, the inftant it arrives upon its ground—the officers move out in front four paces, and divide the ground—the two fergeant coverers replace the two officers in the front rank. The third corporal and the pioneer, cover (in the rear rank) the fergeants—the drummer places himfelf on the right of the right fergeant.

This regards the company when fingle; but when united in the battalion, other pofts are allotted to the drummer and pioneer.

---

Commands { When the Company is to take clofe order from open order.

*Rear ranks clofe to the front* — { The officers, drummer, pioneer, third corporal, face to the right.

March {   The ranks clofe within one pace, marching two and four paces, and then halting.
The officers move into their refpective pofts.
The fergeants fall back, and each individual refumes his place, as in the original clofe order. }

---

The divifion of the company into fquads.

The affembly, and infpection of fquads.

The joining of fquads to form company.

The fizing, and forming the company.

The infpection of the company.

The march of the company by divifions, to the battalion parade.

} Make a part of the internal regulations of the battalion, and require being detailed.

## BATTALION.

WHEN the companies join, and the battalion is formed, there is to be no interval betwixt each, grenadier, light company, or other, but every part of the front of the battalion fhould be equally ftrong—Such intervals prefent fo many flanks, and weak points, and could on no account be allowed in a more extenfive line.

Each company which makes a part of the fame line, and is to act in it, muft be formed three deep.

When the light company is detached and the grenadiers remain, they will be undivided on one flank of the battalion whenever there are feveral battalions in line: but when the battalion is fingle, they are permitted to be occafionally divided on each flank.

*Formation of the battalion.*

*Fig. 5, 6.*

When the grenadier or light company are detached, and make no part of the line, they may be formed two deep, if it is found proper.

---

**Divisions of the battalion.**

The six battalion companies will compose three grand divisions—six companies, or divisions—twelve platoons—twenty-four sub-divisions when necessary for the purposes of march. The battalion is also divided into right and left wings.

The company will draw up according to seniority of captains from the right.

**Position of the companies in battalions.**

Light infantry——lieutenant colonel, second captain——fourth captain, third captain——major, first captain——grenadiers.

The colonel's company takes place only according to the rank of its captain. The three oldest captains are on the right of grand divisions. Officers commanding divisions and platoons are all on the right of their respective ones.

---

There is no interval betwixt companies.

Ranks are at the distance of one pace; except the fourth rank, which has three paces.

**Formation at close order.**

The colours are placed betwixt the third and fourth battalion companies—one in the front rank, and one in the center rank covered by the sergeant major in the rear rank; one other sergeant must also be ready to move out with the front colour, when the battalion marches.

The commanding officer, is the only officer advanced in front three paces.

The lieutenant colonel is behind the center six paces from the third rank.

The major and adjutant are on horseback six paces in the rear of the third and fourth companies.

There is also an officer on the left flank of the battalion.

The pioneers are assembled behind the center, formed two deep, and nine paces from the third rank.

The

The music are three paces behind the pioneers, in a single rank.

The drummers of the six battalion companies, are assembled in two divisions, six paces behind the third rank of the second and fifth companies.

The grenadier and light company drummers and fifers, are six paces behind their respective companies.

The staff of chaplain, surgeon, quarter master, and surgeon's mate, are three paces behind the music.

Every other individual is at his post as already allotted him in company; and in general, officers are to remain posted with their proper companies.

---

In marching and manœuvring the single battalion, the commanding officer is mounted, and in general in the front. In the firings, and when engaged, he is dismounted, and in the center of the front rank before the colours.

The commanding officer of the fourth company may occasionally be ordered in front of the battalion, when marching in line, and may lead it under the direction of the field officers. Commanding officer.

The companies must be equalized, and of course the platoons, this being of infinite importance in all the manœuvres of the battalion and line. In time of service supernumeraries may be collected, or placed to advantage in the fourth rank, ready to fill up the places of such as may be disabled in the other ranks. Companies equalized.

---

| Commands | When the BATTALION takes open order. |
|---|---|
| *Rear rank take open order* - - - | At this *command*—The flank men on the right of the rear ranks, step briskly back to mark the ground on which each rank respectively is to halt, and cover at open distance—Every other individual remains ready to move. |

At

At this *command*, the whole move as follows.

March —
{
The rear ranks fall back two and four paces, each dressing by the right the instant it arrives on the ground.

The officers in the front rank, as also the colours move out four paces. Those in the rear, together with the music move through the intervals left open by the front rank officers, and divide themselves, viz. The captains covering the second file from the right, the lieutenants the second file from the left; and the ensigns opposite the center of their respective companies.

The music form betwixt the colours, and the front rank.

The sergeant coverers move up to the front rank, to preserve the intervals left by the officers.

The pioneers fall back to six paces distance behind the center of the rear rank.

The drummers take the same distance behind their divisions.

The major moves to the right of the line of officers—the adjutant to the left of the front rank.

The staff place themselves on the right of the front rank of the grenadiers, viz. Chaplain —surgeon—quarter master—mate.

The lieutenant colonel and colonel dismounted, advance before the colours two and four paces.

The whole being arrived at their several posts—halt, dress to the right, and the battalion remains formed in order of parade.
}

When

When a battalion is reviewed singly, then in order to make more show—the divisions of drummers may be moved up and formed two deep on each flank of the line—the pioneers may form two deep on the right of the drummers of the right, and the staff may form on the right of the whole.

---

Commands { When the BATTALION resumes close order.

*Rear ranks close to the front* — {
All the officers face to the right.
The music face to the right.
The drummers, if on the flanks, face to the center.
The pioneers, if on the flanks, face to the center.
The staff face to the right.
The colours and lieutenant colonel face to the right.
The serjeants, if in the front rank, face to the right.
}

*March* — {
The rear ranks close within one pace, moving up two and four paces, and then halting.
The officers move through, and into their respective intervals, and each individual arrives at, and places himself properly at his post in close order.
The music marches through the center interval.
The serjeants, drummers, pioneers, &c. &c. resume their places, each as in the original formation of the battalion in close order.
}

---

When the battalion wheels by divisions to either flank into column; both colours always wheel up to the proper front, and place them-

*Colours in column.*

themselves behind the new pivot file.—Drummers, music, &c. remain in the rear of the divisions they covered in line.

---

**No colour reserve.**

There is no colour reserve. There seems no reason why the center should be much more reinforced than any other part of the battalion; and the pioneers, music, &c. sufficiently strengthen it when in order of attack.

---

**Posting of officers.**

All officers are uniformly posted on the right of their platoons. Intricacy arises from the arrangement by wings. On particular occasions, and when necessary, officers are directed to shift to the left of their platoons.

---

**Light company.**

The constant order of the light company when formed in line, and united with the battalion, is at the same distance of files as the battalion. Their extended order is an occasional exception that may be taken when they are single, detached, and when necessary.

---

**Numbering battalions.**

The companies will be numbered from the right of battalions to the left, 1. 2. 3. 4. 5. 6. The platoons of companies will be numbered right and left of each. The subdivisions of companies will be numbered, 1. 2. 3. 4. of each. And these several appellations will be preserved, whether faced to front or rear. The grenadier, and light companies will be numbered separately in the same manner, and with the addition of those distinctions.

---

**Intervals of battalions.**

At all times in line, and with cannon attached, the interval betwixt battalions is twelve paces, being a space sufficient for two battalion guns to work in. Without cannon that interval may be six paces.

With a very few obvious alterations these general rules take place when the company or battalion is permitted, or ordered to form in two ranks only. And it is also evident, that they are meant

equally

equally to apply whether the companies are strong or weak, and whether a greater or lesser number of them compose the battalion.

---

In general the companies should arrive, and halt on the parade of the battalion in open column, either of platoons or companies, and with either *right* or *left* in front. The ground is given by whichever division first arrives at it, and the others arrange themselves in front or rear accordingly.

In this situation are reports made to the commanding officer—companies and divisions equalized—music, drummers, pioneers, &c. assembled at their proper stations—all other individuals of the battalion placed—pivot files, and just wheeling distances corrected. The battalion is then formed in line, by wheels of the quarter circle, and by *word* from the commanding officer, the colours are sent for, and received, and the whole are thus in readiness to move. General method of assembling the battalions.

Sending for the colours.  ⎫
Receiving the colours.     ⎬ Make a part of the internal regulations of the
Returning the colours.     ⎪ battalion, and require be-
&c.                         ⎭ ing detailed.

---

On all occasions of common parade, a battalion or its parts should never assemble, or be dismissed without performing some one operation or other of movement. The line should be repeatedly wheeled to either flank into open column, and from open column into line. When in open column, the several companies should be frequently exercised in the manual and platoon, each by its respective officer; for thereby the men are accustomed amidst surrounding noises to have the eye and ear fixed on, and attentive solely to his command; and officers in the like situation are taught recollection of, and attention to their own business only. Particular divisions should also be wheeled, advanced, or retired, no matter how small a distance, and placed perpendicular to the proposed formation, however little varied from the former one; the others will then be directed to take their places in open column, ready to wheel up into line. The close Mode of instruction.

column may also be formed; and from close column, the line, &c. &c. &c.

<small>Requisites in officers.</small>

In this manner by simple, and imperceptible practice the steadiness, and instruction of every individual is attained; and officers are made masters of the three great and important duties required of them, particularly in the field, and on which so much depends, viz. Precision and energy in their words of command—the judging exactly of distances—the correct dressing of pivot flanks, which ensures the consequent just forming of the line on given, but never on accidental points.

<small>Advantages attained.</small>

The same attentions may on a less scale be had on the parades of the company, or smaller bodies, and on all assemblies of guards and detachments. Was such method uniformly practised, much time unnecessarily consumed in the field in detail and manual exercise might be saved, and the battalion be there solely employed in movements, arising from such circumstances and varieties of ground as presented themselves to the commanding officer. The modes of their execution, would be already thoroughly understood, and instantly applied by each individual.

---

<small>Fig. 7.

Attentions in the exercise of small bodies.</small>

Single divisions, companies or battalions when at exercise, must generally consider themselves as part of a line, and not always as detached, or independent bodies. Their movements, and formations should be on a supposition of lining with other troops, already upon their flanks. Two or more persons separated at a proper distance from one another, and from the company or battalion, may represent the flanks and center of an adjoining battalion, and may always first take their station in a new line—This would cause the formations to be on determined, not accidental points; the universal practice of which latter usage is, what in part occasions the incorrectness and deficiency too apparent, whenever any number of our battalions are directed to move, act, or form in concert.

The idea of a battalion being confidered as a perfect and independent body, and its commander playing the part of a general on moſt occaſions is very apt to miſlead. In war though it may be at times ſingle and detached, yet in common it ſhould only be looked upon as a member of the line, its movements as relative and dependent on thoſe of others, and its principal operations ſhould be calculated accordingly.

In order to enforce theſe eſſential obſervances, general officers at reviews, and inſpections ſhould not allow regiments to move, and form, but in ſuch given lines, and to ſuch given points, as they themſelves ſhall at the inſtant preſcribe—they ſhould narrowly obſerve that commanding officers are regulated by theſe principles, and do move, and form on two detached objects in the ſame ſtraight line, not on one alone, which is the general practice, and which muſt always produce an accidental march or formation of the line.

<sub>Attentions of general officers at reviews.</sub>

OF THE

# C O L U M N.

A COLUMN, is any number of separate bodies placed in a continued line behind, and covering each other.

When there are distances betwixt each, it is called an open column, or column of march, or manœuvre.

When there are no distances, but the several bodies composing it are closed up to each other, it is called the close column, the mass, or column of formation.

Fig. 8.

---

All columns are supposed formed from line, and for the purpose of again extending into line—either flank may lead, but the divisions of the line must follow in a regular manner, from right to left, or from left to right. Whatever is the relative position of a body in line, such ought it to be in column; and where several connected columns are formed, the same flanks of each should be in front; but whether rights or lefts, will depend on circumstances.

General method of forming columns.

The advantage attending this column, is, that it is universal, simple, and best adapted to all future formations of the line, by one easy, general, and constant method; and that at any one instant, and upon any one given division, it can extend into line. When its mechanism is considered and well understood, its formations

mations will all appear relatively *central*, made on the shortest lines, and in the quickest manner possible.

Columns formed from the center of the bodies, which compose them by alternate divisions, files, or any otherwise composed than from the flanks; are for the most part defective and improper, not general or adapted to all situations, having intricate methods of forming in line, which vary according to circumstances; and should not be used, as being unnecessary, and tending to puzzle and embarrass at times, when precision is most requisite.

<small>General method of forming columns.</small>

The column formed from the flanks, is adapted to all possible occasions of movement; and is readily converted into that of attack, or defence. But, at the same time it is allowed, that there may be some given situations, where the column may be formed to advantage from the center.

## Of the OPEN COLUMN.

THE great object of manœuvre, is to be able from the order of march, to take in the quickest and best manner possible, any given position in line; and instantly to change from that situation, to any other, which circumstances may require, and which will give an advantage in attacking, or opposing an enemy.

<small>Necessity of making changes of situation in columns.</small>

This can seldom be done by the movement of large bodies in front, which will always take up much time, is fatiguing to the troops, and difficult in proportion to the extent of such front.

Changes of situation in great bodies, from one distant point to another, are therefore generally made in one or more open columns.

When such columns are arrived near their ground, and the line is again to be formed in any given position: as the irregularities and obstacles of the surface, will often not allow the divisions to march up in front, a method general, and adapted to all situations must be fixed upon; for this purpose, the filings of divisions from a flank, are particularly applicable to the internal movements of each battalion, as enabling in all varieties of ground to take the shortest lines, and to make the formation in the quickest manner possible. *General mode of forming from column to line.*

But on no other occasion, ought the battalion or larger bodies, ever to change their situation in file—all marches and movements must be accomplished, in column of companies, platoons, or subdivisions in front; those columns multiplied according to circumstances, and the most scrupulous exactness required in the prescribed distances; on the judging and preserving of which, all military manœuvres depend.

---

The movements of the OPEN COLUMN, are the foundation and mean of all route marching; and of all transitions from one position to another.

It assumes the name of column of march, when applied to common marches, where the attention of men and officers need not be kept on the stretch. *Column of march.*

It assumes the name of column of manœuvre, when being within reach of the enemy, the greatest exactness is required, in order to its speedy formation into line; or to the change of situation of that line, from one position to another. *Column of manœuvre.*

The chief objects of the open column, are, facility of movement, the quick formation of the line to the flank, and the change of situation in the shortest lines, from one position to another. *Objects of the open column.*

Columns of march or manœuvre, will generally be composed of companies or platoons; except when in column of march, the narrowness of the ground makes it necessary to advance on a smaller front. *Front of column.*

The commander of the column is always in front of it. *Commander of column.*

In column, officers are three paces in front of their divisions, when the continuation of the march is the object, and the flank non-commissioned officers then preserve distances: but as soon as *Officers in column.*

they

they enter the *alignement*, and the intention is to form in line, they then place themselves on the pivot flanks of their divisions.

---

Distances in column.

The distances in column, are always taken from the flank man of the front rank, to whichever hand the column is dressing. The pivot officers and men, take care to cover exactly in a line from front to rear of the column; so that if the distances are just, the line may at any time be formed to the flank by wheeling up of divisions.

Distances of divisions.

The distances wheeled at from line, are always preserved in open column, unless ordered occasionally to close them.

Column at half intervals.

Where the formation of the line to the front or rear is intended, and that to the flanks cannot be required; the column may march at half or quarter intervals of divisions, and form; either by entering into the alignement, taking distances, and wheeling up to the flank; or it will move up into close column, and then deploy into line.

Intervals in column.

When in march, the proper division will preserve the distance of intervals (when any) in addition to that of its own front.

Dressing in column.

In column, the whole divisions dress and cover to the proper pivot flank, viz. to the left, if the line has broke to the right; and to the right, if the line has broke to the left.

Distance of ranks.

Rear ranks are one pace asunder. When the column has to march a considerable distance, they may be allowed to take two paces, but without increasing the distance betwixt divisions, which remain such as are prescribed according to the object of the march or movement.

Orders how given in column.

In open column of any kind, the words of command from the chief leader, are only addressed to the leaders of battalions and divisions, and not to the men: such leaders give and repeat to their respective bodies, the executive words of command, which must circulate rapidly from front to rear of the column, and be given with distinctness and a loud voice.

Orders given in column.

The divisions of columns, may receive orders from the commanding officer of such divisions, whether they are looking to or
from

from such officer. In general, he is on the flank to which the men dress, always leads when in file, and conducts the pivot flank when marching in the alignement.

---

The open column will never exceed such a pace, as the rear cannot readily comply with, without hurrying. But when the head of the column has at any time changed its direction, should the rear be ordered to file; that it may the sooner gain such direction, the head must then move at a slow pace.  March in column.

In some situations, the different divisions of the column may necessarily be marching at different steps.

All wheelings and filings of divisions made from the halt from line to form in column, or from column to form in line, must be at a quick step.

In filing, the leader of each file is attentive to his point of marching, his interval from wherever he takes it, and to dressing the head of his file. The followers of the front rank cover the necks of their leaders exactly, and observe distances. The men of the rear rank look to, and are regulated by their proper file leaders of the front rank.  Filing.

In division filing for any distance, the files may be loosened and again closed, before they arrive at their forming point.

---

In open column of march or manœuvre, the artillery, music, drummers, &c. wheel with and remain closed up to the rear, or are on the flank of the divisions to which they belong when in line.  Artillery, music, drummers, &c.

In column at half or quarter distance or in close column, they will file on the flank, which is not the pivot one.

In the square or oblong, they will be in the interior part of it, or follow it when marching.

To prevent an unnecessary inversion of the line, columns of march or manœuvre, should be formed with the left in front, whenever it is probable that the formation of the line will be required  Inversion of the line.

to the right flank, or that the enemy is on the right of the march; and *vice verſa*, when on the left of the march.

The column when marching at half or quarter intervals, muſt preſerve a diſtance betwixt battalions, equal to the front of the column.

## GENERAL FORMATIONS of the OPEN COLUMN.

Fig. { 11. 12. 13.

Formation of the open column from line.

From line, the open column is formed, and marches to the front, flanks, rear, or in any intermediate oblique direction, with either its right or left in front.

In each caſe, the battalion or line *wheels* the quarter circle by diviſions to either flank, and halts—the whole *march*—the leading diviſion *wheels* into, or moves on in the preſcribed direction, and the reſt follow in column.

Fig. 9.

The open column, or the column at half or quarter diſtance, is alſo formed penpendicular or oblique to the line, in any given direction, and on any given diviſion. By the other diviſions (according to which flank is ordered to lead) wheeling, filing, and placing themſelves in front and rear of the given one.

Formation of the line from open column.

The open column, will form in line to the front flanks or rear, and in any oblique or perpendicular direction.

In all caſes, the column moving from its original ſituation, will by its various operations, arrive, and place its pivot flanks in the new intended direction; it will then wheel into line.

## The BATTALION in OPEN COLUMN.

HAVING mentioned the general circumstances that relate to the open column, it becomes necessary to ascertain more minutely the operations of the battalion when in that situation, and during its several transitions from line to column, or from column to line. The column is here generally understood, as being formed with the right in front; but when it is a column with the left in front, the necessary alteration of commands and movements are evident, and easily made.

### When the BATTALION breaks to the FLANK into COLUMN.

THE commanding officer of the battalion orders: by *platoons* to the right (or left) *wheel—march*. All the platoons wheel at the same time, and observe what has been already prescribed in the wheeling of a single division—when the leading platoon has compleated its wheel of the quarter circle, it receives from its own commander, the words: *halt—dress*. The others receive the same words, and in the same moment, which must of course happen when all the platoons wheel with the same quickness and length of pace.

When the battalion is thus broke, and halted in column, the commanding officer gives the word: *march*, the whole then step off together and proceed. The pivot leaders of the two front platoons will move steadily on the points of march given them, and the following ones will cover.

*[Margin: When the battalion formed and halted is to break by divisions, and march in column from one of its flanks.*

*Fig. 13.]*

---

When a battalion from line is to break to either flank, an under officer belonging to the leading platoon, previously runs out to mark the point, at which the pivot flank of that platoon is to halt after the wheel: an over wheel is thereby

thereby prevented, and the other platoons are also better directed.

<small>Attentions when the battalion breaks into column.</small>

The officers wheel with their platoons, and after dressing them, place themselves three paces before the center of each, where they march till they enter the alignement; they frequently look behind to observe their platoons, and to preserve the same pace—the men dress to the pivot flank, and the non-commissioned covering officer who has shifted to that flank, is particularly attentive to preserve the distances of platoons.

Should any of the platoons be stronger than the others, and out-flank after the wheel; such platoons must cover or be covered, on the new pivot flank by the weaker ones, and such as are ordered will shift accordingly.

When the battalion or line breaks into column, both colours will wheel up, and remain behind the pivot flank of the leading center platoon.

---

When the COLUMN enters and marches on the ALIGNEMENT.

<small>When a battalion marching in column, enters into the alignement and marches along it.</small>

A battalion marching and broke from the right, may enter the alignement by wheels, either to the right or left. In either case, the left or pivot flank of the platoons must be placed on it: in the first instance, behind it, and in the second, before it. In both cases, the line is afterwards formed by wheels of platoons to the left; in the first instance, the line will front the same way as the column; in the second, it will front to the rear of the column.

---

Let us suppose then, that the given points of formation are A B; that C is the point of wheeling; D the adjutant already placed;

placed; that the battalion hath marched from the right, and enters the alignement by wheels to the right.

The pivot flanks of the column, having been directed ftraight on the point of wheeling C; when the leading officer has arrived at a diftance equal to the front of his platoon from it, he orders: *halt*—to the right *wheel!* and then if he has preferved juft diftance, his left wheeling man will nearly touch the point C, and arrive on the alignement—without delay; and after he has ordered *halt— drefs!* he will place himfelf on the left flank, and order: *eyes* to the left—*march!* he does not ftep at any of thefe commands, but they fucceed juft quick enough to allow time for their execution, and to avoid interrupting the officer that follows him. After this he moves on without looking behind, or allowing any thing to take off his attention, and at the eftablifhed ordinary pace towards the diftant point; fo that his perfon fhall juft graze the head of the adjutant's horfe, which he invariably preferves in a ftraight line with the point B. This rule, all the following officers muft obferve; and fhould any of the platoons deviate to either hand, thofe that fucceed them muft rectify the fault, and exactly touch the point where the adjutant is placed.

Fig. 15.

---

The principal attention of the leading officer muft be, never to change the time or length of ftep, otherwife, a ftop muft happen in a confiderable column, and the foldiers will afterwards be obliged to run. He muft march in one conftant pofition, and take particular care that his front rank is perpendicular to the line on which he marches.

The fame directions regard the other officers who conduct platoons, and who in addition muft correctly obferve——— That at the word *march*, given to the preceding platoons, the following one is ordered: *halt—wheel*. In this they will exactly agree, if the officers preferve their due diftances, and make their wheels at a redoubled pace—and alfo, that all the platoons wheel at the identical point, where the leading one wheeled; therefore, all the platoons muft march ftraight

Attentions when marching in the alignement.

straight on the firſt rank of the preceding one, which hath already wheeled. It commonly happens, that the platoons of a battalion in wheeling to the right, incline to the left from the original point ſucceſſively; a fault which produces a defective march in the line, and ſhould be prevented.

<small>Attentions when marching in the alignement.</small>

Pivot officers of columns, when marching in the alignement, muſt be ſteady on the flanks of their diviſions: as they give the true wheeling diſtance and covering of the pivots in their own perſons, they muſt not look to, or endeavour to correct their diviſions; their attention is otherwiſe ſufficiently employed, and that care muſt be left to the ſergeants, and other officers. The pivot files of men muſt be quite cloſe to their officers, that they alſo may be truly covered when halted; the colours are behind, and in the line of thoſe files of men, nor muſt there be at any time more than one officer's file on each pivot flank.

All marching in the alignement muſt be made in ordinary time, and taken up correctly from the point where it is entered——All doubling up, or increaſing the front of the column of march, muſt be made before entering it.

---

When a battalion broke to the right, enters the alignement by wheeling to the left; the leading platoon begins its wheel to the left on the alignement itſelf, when the pivot flank arrives at the point of wheeling; inſtead of (as in the foregoing caſe) being removed the diſtance of a platoon from it.

Whatever has been ſaid reſpecting a battalion broke from the right, takes place in one broke from the left: the only difference is, that the flanks are now changed; that the left platoon does, what before was done by the right; and that the right flanks are placed on the alignement, inſtead of the left.

---

To inſure the more correct march and halt of the pivot flanks, in the alignement—the leader of the column may occaſionally go forward

forward to the advanced adjutant; and being himself truly placed, may look back to the point of wheeling or entry into the alignement (if there is no other more remarkable object in it). He can then see if the rear flanks of the column keep the true line or deviate from it, and may correct them by signal, or by sending back an adjutant to take his position in the true line, and to whose direction they are immediately to conform.

In this manner also can the leader if necessary, correct the pivot flanks after a halt, when there is a rear point of view sufficiently marked—if that is not the case, he may go towards the rear of the column, line the flank of the fifth or sixth platoon on that of his leading platoon, and the front point of march; he will then return to the first platoon, and on the flanks of that and the fifth, correct the rest of the pivots.

<small>Pivot flank.</small>

---

### When the Column HALTS—FORMS in LINE, and DRESSES.

WHEN in the manner above prescribed, one or more of the battalions are marching in the alignement there can be nothing easier, than to form well in line. On the word or signal being given: the commanders of battalions order *halt*—this must be pointedly complied with at all times, even though there should be false distances betwixt the battalions. No officer moves after hearing the *halt* of his commander not even a half pace, but the foot which is then off the ground finishes its proper step, and the other is brought up to it— If that was not done, and that one officer should stop while another was permitted to make one or two paces, those behind would be obliged a-new to shift and great confusion would arise, from officers being

<small>When a battalion marching in the alignement, halts and wheels to the flank by platoons, to form in line.</small>

<small>Fig. 14.</small>

being deficient in one great principle of their bufinefs, the preferving of proper diftances.

<small>Halt in the alignement.</small>
Whenever the *halt* is ordered; the more eafily to adjuft the pivots, the officer (and covering under officer) falls behind the pivot file of men, and examines that his files are well dreffed by the pivot flank, and places himfelf on the right of his platoon. If the wheel is to the left, he afterwards wheels up with his platoon: If the wheel is to the right, he rather keeps back, allows it to be made on the pivot man, and does not move into the line till the wheel is compleated.

---

<small>Form in the alignement.</small>
When the word is given *platoons wheel* (to the right or left) —An under officer of the leading platoon gives ground to the flank of that platoon when it wheels up into line, in the fame manner as when it wheeled into column: he advances, lines himfelf correctly on the pivot men, and exactly afcertains the point where it is to halt. At the word *march*—All the pivot men ranged in the alignement face into the new line, and remain immoveable on their ground; and the oppofite flanks wheel up to join the pivot men of the preceding platoons who are thus already placed. The word *halt* is given to each platoon as it arrives in line.

It would appear to facilitate much the dreffing of the platoons, if the pivot men inftead of facing at the word *march*, fhould remain immoveable and not front, but by order of the commander after the battalion is dreffed; as the officers could then better diftinguifh thofe fixed points, than when they front the fame way as the reft of the platoon which are dreffing.

---

<small>Drefs in the alignement.</small>
At the word *drefs*—The officers move out quick to drefs their platoons in the manner already directed. As there are fo many determined pivot points given, it becomes eafy to drefs correctly a platoon or a battalion; and great care muft be taken that the pivot men do on no account move up, or fall back whatever directions may be then giving for compleating the dreffing. If a defect exifts, it muft proceed from the other men not having lined with thofe

fixed

fixed points. The internal corrections of platoons must therefore be made, but the original pivot men should remain immoveable.

The officer of the third platoon for example, has only to consider the left file leader of the second platoon (close to whom he stands) as the *point d'appui*, and his own left flank man as the point to dress upon, there will then be nothing easier than to dress his platoon without moving from the spot. But he will still more exactly do it, if he places himself 2. or 3. files on the other side of the flank man of the second platoon.

If all officers are in this alert, and skilful, and that the soldiers are accustomed to correct themselves, a battalion will instantly be formed, nor will the commander or major have any thing to rectify. Should it be objected, that the battalion will be ill dressed if the pivot flank men have not covered each other well in the alignement: it is answered, that the platoons must never wheel up till that is done; and as there is seldom time to correct such faults, the greater care is to be taken not to commit them.

Attentions of officers in dressing.

---

The point where the head of the column enters the alignement, which is given to, and never must be quitted by a mounted officer, but as he is relieved, and till the whole have entered it—The several adjutants who place themselves in the true line—the point where the flank of the leading battalion begins to form—the colours of battalions which have truly halted, or are formed up——these are so many marked points within the line itself independent of distant objects, on which the covering of pivots, or the formation and dressing of battalions can be regulated.

The alignement of the march will always be about a pace before the line on which the troops form: because the one is the direction preserved by the officer in marching, and the other being that on which the flank men halt, and the platoons wheel up into line, is distant of course from the other the breadth of a file. Although the officers halt in the true alignement, yet it is impossible to allow them to remain there immoveable as points of forming for their divisions, becaufe the dreffing of thofe divifions depends on them. The flank files of men are therefore the pivots of platoons in wheeling up into line although the officers are the pivots during the march, and an attempt to form in line exactly on the points of march would derange the pivot files of men and caufe diforder.

A commander fhould drefs as little as poffible from the center, at leaft he fhould be careful that he himfelf is in the alignement, which is alfo a neceffary caution whenever he directs his platoons marching in the alignement. He muft not as often happens, pufh his horfe betwixt two platoons, for he himfelf will not then be lined, and it will occafion the officers to lofe the alignement. If he will fee whether the platoons march well, he muft place himfelf on the alignement, and on the adjutants who are in it, and give his horfe fuch a direction as the platoons fhould touch in marching.

---

When the left of the column is in front.

When broke to the left, all that has been faid takes place, and is in the fame manner executed—Only—the right flank man does what has been directed for the left: he fronts, when the platoons begin to wheel up; and the point d'appui being now on the left, the dreffing muft from thence be regulated; confequently the foldiers look to the left. But as the officer in order to aligne his platoon is obliged in fuch cafe to fhift from right to left, which muft occafion a delay, it would be better that the feveral officers fhould drefs the platoons which are on their right.——In this manner the flank officer would drefs the left platoon, &c.—the officer of the feventh platoon will place himfelf to the left of the right flank man of that platoon and drefs the fixth on its own right flank man—the fixth officer will drefs the fifth platoon, and fo on, till the fecond officer dreffes the right platoon.——In this manner will a battalion be fooner dreffed, than when each officer alignes his proper platoon.

### When the Rear of the Column muſt File to enter the Alignement.

It has been hitherto ſuppoſed, that the whole platoons of a battalion have entered into the alignement before the halt and the conſequent formation into line is ordered. But as it will often happen, that the head of the column muſt enter not far diſtant from the point where it is to halt and be placed, and when it arrives there, that the rear platoons cannot at that time have alſo entered, but are ſtopped in the old direction by the ceſſation of movement in the front; it becomes neceſſary immediately to bring thoſe platoons into the alignement, that the battalion may juſtly form.

*When the leading platoons of a battalion halt at their point of alignement before the rear ones have entered it.*

Whenever therefore the leading platoon of a battalion arrives at the point where it is to form into line, and that this is the object, the word *halt* is given, and the whole ſtop. The leading platoon and ſuch as may have already wheeled into the alignement being now at their proper points, remain ſo; and the word *face* (to the right or left as neceſſary) is then given, when all the platoons who are ſtill in the old direction face to the flank. At the word *march*, each platoon moves quick in file in a ſtraight line towards its point in the new line. As ſoon as each platoon arrives perpendicular with its pivot flank and at its wheeling diſtance on the alignement, it receives ſeparately the word *halt*. The word *front* is then given to each platoon; and when the rear one has taken its place, the whole battalion ſtands in column on the alignement, ready to wheel into line when ordered.

*Fig. 18.*

---

When the column is marching in front, and the leading platoon by a *wheel* gives a new direction and moves on—If the object is a continuation of the march, the other diviſions ſucceſſively *wheel*, follow its exact tract, and when the head *halts*, the rear will alſo halt in the ſituation it then finds itſelf in—If the intention is to

*When the column wheels into a new direction, and continues its march.*

**Fig. 15.** form in line, the rear will then be ordered to *file* into the new direction: but should the column proceed in its march, the rear divisions will continue to follow.

---

### When the Column is to form Line on a Central or Rear Alignement.

**Change of position of the column to form line on a central or rear alignement.**

The Alignement or line of formation hath hitherto been supposed in front, and that the column hath advanced to enter into it. But there may be cases where the column has overpassed it, or where for other reasons a formation may be required which is either in the rear of the column, or which intersects it.

---

**On a rear alignement.**
**Fig. 17.**

When the alignement is in the rear of the column, each platoon will separately countermarch. The column then having changed its front, will advance, enter the alignement, and be prepared to form according to circumstances and the modes already prescribed.

---

**On a central alignement.**

When the alignement intersects the column—The platoon which it cuts must be placed perpendicular to, and with its pivot flank upon it, fronting whichever way the column is to extend. The platoons which were ranged before the given one, must now be considered as in the situation of a column with the alignement in its rear, they will therefore separately countermarch and face it; by this means the alignement becomes in front of each of the two component parts of the column, and they can arrive in it by the same identical operation. For this purpose the whole platoons, except the central one which stands fast, file to whichever (but the same) flank is on such occasion proper, and place themselves as before in column by halting and dressing their pivot flanks in the new alignement. Such as now face the given platoon will again be ordered to countermarch and be in its front.

( 85 )

When the column is thus formed, it is ready to wheel up into line, to march to the front, or by firſt countermarching the platoons to move to the rear.

In a column compoſed of ſeveral battalions, there is a neceſſity that it ſhould follow the above preſcribed method, and that all the platoons of it that are in the front of the central one, ſhould each *countermarch* and *face* it, before the change of ſituation can take place. Becauſe, the platoons of the central battalion alone file, but all the other battalions *march*, each in column of platoons from their inward flanks to the new alignement, and eaſily enter upon it taking their diſtances from the front. Were thoſe platoons not to countermarch, the outward flanks of the battalion muſt lead and firſt enter on the alignement; and all diſtances muſt in that caſe be taken from the rear, which would be a very difficult and uncertain operation, where ſeveral battalions had to enter the line. *Reaſon of the countermarch in a conſiderable column. Fig. 20.*

But undoubtedly a ſingle battalion in column, or any body the whole platoons of which file, may change its ſituation, without the previous or after operation of the countermarch. When the central diviſion is placed as before, the rear platoons will file up, and the front platoons will file back, and arrange their pivot flanks in the alignemnt; in doing this, the rear diviſions have every facility of moving up and lining with the fixed points in their front; but the platoons before the central one, muſt take their diſtances and line from the rear. *Countermarch may be omitted in a ſingle battalion.*

---

In general, the line is not formed till the whole of a battalion is arrived at its ground, halted, and adjuſted. It then is ordered to form by each diviſion wheeling the quarter circle, and thus battalion after battalion.

But if neceſſary, when the leading diviſion of a battalion has got to its point, it may wheel up and form in the new direction of the line, though the others are not yet arrived in it; and thus diviſion *Wheeling up from column into line.*

Fig. 19.

division after division successively as they come up. In this manner however, the exact dressing of the battalion is not so soon or so easily obtained, as if the pivots of the whole battalion are first accurately lined, and then that the divisions wheel up and form. If part of a battalion should wheel into line, while the other divisions are coming up into column; the pivot men of those divisions, and not the officers, must cover in the formed part of the line.

## Countermarch of the Column.

Countermarch of divisions to change the front of the column.

The countermarch of divisions is equivalent to a wheel to the right about, but performed with more convenience in much less time, and without change of ground. That of each division separately when in column, is an evolution of great utility on many occasions, and enables a column which has its right in front, and is marching in the alignement, to return along that same line, by becoming a column with its left in front, and to take such new position in it, as circumstances may require, without inverting or altering the proper front of the line. In many situations of forming from column into line, it becomes a necessary previous operation.

Fig. 20.

As soon as the commander has ordered the column to *halt*, and to *countermarch* by divisions, each division separately receives from its own officer the words—to *face—march—halt—front*; and the column is then formed to its former rear, ready to move on when ordered.

The countermarch of divisions in column, is generally made from the right of each, and behind the rear rank of each. In this operation, besides changing its front, the division must necessarily shift its ground a space, equal to twice its depth.

When

When it is neceſſary that a battalion advancing in column, which would naturally form to the left, ſhall be prepared to form to the right, without inverting the order of the battalion—A countermarch of the whole column by ſucceſſive wheels of the diviſions to the right about muſt take place, and then a countermarch of each diviſion of the column; or elſe a movement equivalent thereto, which changes the wings, and brings the one which was before in the rear, to the front of the column, without, at the ſame time, preventing the continuation of its march. Countermarch of the wings to change the flanks of the column.

When therefore the right is in front, that the left is to be brought up, and that there is ſtill ground for the column to advance; the commander orders: *halt*—the *left wing* to the *front*. The officer of the left platoon immediately orders it to *face* to the right—and *march*, till his left flank can freely paſs near the right flank of the others. He then commands: *halt—front*, and moves on quick and cloſe by the right flank of the platoon, then preceding him. The officer commanding that platoon, as ſoon as the other approaches him, commands: *right, face—march*, behind the now leading one—*front*; when he covers him, and then follows at his due wheeling diſtance. All the other platoons, ſucceſſively, perform the ſame operation; and when the right platoon has taken its place in the rear, the column reſumes the ordinary ſtep. If before this operation, the column is cloſed to half or quarter diſtance, then all the platoons *face* at the ſame time, proceed as above directed, and each takes its diſtance from its preceding one before it moves on. Firſt method.<br>Fig. 21.

Another manner is when the front is to be changed on the ſame ground, ſo that the left wing comes to the ſpot which the right wing occupied, and ſo *vice verſa*. Second method.<br>Fig. 22.

In this caſe, the left platoon proceeds exactly as has been already directed; all the others go to the right about, and march on at

a quick

a quick step towards the place from whence the left moved. When the platoon next it arrives at that place, it receives the order to *face* to the left, passes behind the left platoon, *fronts*, and follows it: and in this manner, all the rest proceed till the right platoon when it fronts, finds itself where the left hath been; only, that the whole column is removed to the right, a space equal to its front.

It must be observed as a general principle, that the platoons which advance, come always out on the side to which front is to be made, and on which the enemy is placed; because then with the platoons which are free, he can be opposed while the others are moving behind the line.

In a column composed of several battalions, where an inversion of the battalions within themselves, but not of the wings, is meant to be prevented; then each battalion separately will perform this operation. But if the inversion of the wings also is to be avoided, then the whole column will proceed as if it was a single battalion.

---

Third method.

Fig. 22.

There is a third method by which the alignment is in no respect altered, it may be used near to or on a parade, but not where there is any thing to be apprehended from an enemy.

The column standing marched from the right, should naturally form to the left, but it is here meant to form to the right; therefore all the platoons except the last, at the word: to the *right* and *left open*; by the side step do open half to each flank, a space sufficient to allow a platoon to march through in front. The left platoon does not open, but passes through the others at a quick step; and as soon as its rear rank arrives at the front rank of the one next it, that platoon *closes* by the side step, and follows; in this manner, they succeed each other till the column is formed as marched off from the left, and then resumes the ordinary step.

But if the ground of the column is not to be changed after opening out, the last platoon moves on, the others face about, march

march quick succeffively, face inwards and join, front, and follow till the flanks are changed; and that the left platoon halts exactly on the ground on which the right stood.

---

A battalion already formed in line, may change its front by a countermarch from both flanks on the center. The right files move close behind the rear rank, the left before the front rank of the battalion, till they arrive at the points where each other respectively stood; they then front and dress to the colours, which have kept their ground, and served as the pivot on which the battalion turns. *Countermarch of the battalion by files on the center.*

---

### Taking Distances from the Rear.

It has been shown how the column changes its front by the countermarch of platoons, and returns towards any new point D'Appui. But situations also occur, in which it is necessary to take distances from the rear of the march, in consequence of the last platoon being ordered to stop at a given point, and thereby regulating the halt of the others, which in such case must proceed from rear to front, either instantaneously or successively, according as the column is marching, at whole or lesser intervals. *When the rear of the column is to give the point of appui, to take distances from it.*

---

The column at quarter or close distance, and with its right in front, hath halted at A, with a view to deploy to the right, on the supposition that the enemy is formed in B; but finding that he hath changed his position, and is actually in C, it becomes necessary to give up the first intention, which in the position D would present the left flank to the enemy, and to take another in E or F, so that the left wing shall still remain posted on the height *Fig. 23.*

height A, but without expofing its flank. In fuch cafe, were the diftances to be taken from the front, either the height or the intervals betwixt the battalions might be loft, if any one officer fhould err in his diftance, or if the commander fhould order *halt* too late. It is therefore neceffary they fhould be taken from the rear.

<small>When diftances are taken from the rear of the column.</small>

The column being apprized that diftances are to be taken from the rear, the whole march off at the quick ftep, having previoufly placed adjutants in the alignement to give diftance to the leading officer, who as well as all the others is on the pivot flank. When the laft platoon of the column arrives at A, chofen as the point d'appui, it is ordered to *halt*. The officer of the platoon who precedes it, looks back, takes for his flank man, as much diftance as is neceffary for the platoon to wheel up in, and then *halts* it. In the fame manner, the officers of the other platoons proceed quite to the front.

When the platoons are all placed, the line is formed by a wheel to the left, and dreffing is of courfe taken from the point d'appui which is on the left. In general, no officer will miftake the hand to drefs to, if he will attend to his point d'appui and recollect that it is from thence, without exception, that he muft drefs towards another diftant point.

But as the officers cannot with fufficient exactnefs, enter into the alignement, on account of the quicknefs of the march, and becaufe they are obliged to look behind for the diftances; the obfervance of the following rules will correct any fubfifting faults.

<small>Attentions in placing pivot men.</small>

The officer of the laft platoon, who is pofted at the point d'appui, will drefs the pivot man of the preceding platoon, on the adjutants and advanced point of march. As foon as this man is placed and fteadied (by fignal from the officer of the rear platoon), his officer looking upon himfelf as a new point d'appui, in the fame manner lines and halts the pivot man of the preceding platoon, and thus fucceffively till it reaches the front. It is not an unneceffary precaution to divide the alignement by adjutants, as many

obftacles

obstacles may prevent the officers seeing the distant points in it.

## The LINE in OPEN COLUMN.

HAVING now detailed the movements of the battalion column, it is necessary to apply them to those of more considerable columns, as their operations are directed by the same general rules.

### Of the GENERAL MARCH in COLUMN.

NOTHING so much fatigues the troops in a considerable column, and is more to be avoided than an inequality of march. The principal reason of this is, that the rear of the column frequently and unnecessarily deviates from the straight line, which its head traces out, and moves in a circular or serpentine one. In endeavouring occasionally to regain that line, and their just distances, the platoons must of course run or stop, and again take up their march; for if they preserve their just distances when marching on a curved line, they cannot have them when brought into the straight line drawn between its two extremities.

*Causes of irregularity in the march of the column.*

The remedy to this is, a greater exactness in covering the pivot leaders; and that whether the officer or under officer is conducting the flank of the platoon, he should always preserve two or more pivot men in the straight line before him. It is unnecessary to attempt the same scrupulous observance in common route marching, as when going to enter into the alignement; but even a general attention to this circumstance, will in that case, prevent unnecessary winding in the march, which much prolongs it and fatigues the soldier.

## Of the MARCH in the ALIGNEMENT.

*Attentions necessary to the march of a considerable column in the alignement.*

THE same rules that direct the entry and march of a battalion in the alignement, apply to those of the most considerable column. But in addition, there are also certain observances of the staff officers of the battalions, to place and preserve each in that alignement; and their aid is more particularly required, where the inequality of the ground in heights and valleys, prevents the marked and strong points in the direction from being seen.

Fig. 25.

The commander or major of each battalion, place themselves successively as has been prescribed, at the wheeling point, and determine the just entry of their battalions into the alignement. The pivot flanks of the preceding battalions become so many points of direction, in addition to those furnished by the adjutants.

As the leading battalion will be most apt to deviate from the true line; the majors by occasionally lining themselves backwards on the adjutants and the point of wheeling, can correct such inaccuracies; and when the troops are in low ground or crossing a valley, they can in this manner direct the march at times, when the front points of view are not seen. In such situations, adjutants must be placed both on the heights, and also in the intervening valley, in order exactly to prolong the alignement.

The general line is always to be followed, not the partial deviations of a preceding battalion, which if it swerves or loses its true alignement, must by the attention of the staff officers immediately regain it.

FORMATION

## FORMATION of the LINE from COLUMN.

WHEN a confiderable column is to halt in the alignement; in order not to lofe diftances, the moft inftantaneous ftop of the whole body fhould be made at the word *halt* from the commander of each battalion, repeated fharp and without the leaft paufe or interval.

*Attentions neceffary for the formation of a confiderable column into line.*

As the march in the alignement ought always to be at the ordinary pace, and as the ftep is fuch as allows of its being occafionally lengthned, the fmall inaccuracies of diftance may be corrected during that movement; therefore, unlefs there is great inattention at the wheeling point, and afterwards in the feveral ftaff and platoon officers, in preferving the pivot lines and wheeling diftances, the column on halting as above ought to be correctly dreffed, and ready to wheel up into line.

---

Should the column be required to march in the alignement at a quick ftep; although the pivot line may be correctly obferved, yet juft wheeling diftances cannot be expected to be preferved, and the intervals between battalions will probably be increafed. But if thofe between the platoons of each battalion, can be kept nearly juft, each battalion may fucceffively *halt* when its leader fees it, at its due diftance from the preceding one.

*If a quick march is required from a column.*

---

When a confiderable open column, or columns at half or quarter diftance, arrives near and behind any part of the ground where it is to form, it will be halted. The commander will determine the exact direction of the new line, and where his leading flank is to be placed; officers will then be detached to afcertain flank points to their feveral battalions.

*Arrival of a column of feveral battalions in the alignement.*

The

( 94 )

Fig. 31.

The front platoon of each will then be directed towards its point, the remainder of such battalion will follow in column; and when the front division of each is arrived at its ground, and is *halted* perpendicular to the alignement, the other platoons will be ordered to *file*, and arrange themselves anew in column behind it. The several battalions will form separately in line, by the divisions of each wheeling the quarter circle.

Officers that lead battalion columns into a new line, must take great care neither to overshoot the line, nor to crowd upon the battalion that precedes them——As soon as the leading division of a battalion column is at its ground, a mounted officer must from the pivot of that division, correct and dress all the other pivots of this battalion on the given points, whether in his front or rear, on which the line is to be formed; and when so corrected, the battalion is ready to wheel up into line.

In whatever manner the leading flank of a battalion arrives in a determined line, a mounted officer always gives the precise point where it enters, and does not quit it, before the whole are in that line.

---

Manner in which the rear of the column enters the alignement when the head halts in it.

Fig. 24.
Fig. 26.

Should the column enter on one flank of the line, it will continue its march till the whole are entered, and till the head halts at the other flank.

But should it enter at any intermediate point, the leading platoon when it arrives at such point, will *wheel* and continue its march in the alignement, till the column is ordered to *halt*; the rest of the platoons which have continued to follow, and are now in the alignement, will remain so at their just wheeling distances—And such platoons of the battalion that have last begun to enter the new direction, but are still in that of the old column, will immediately be ordered to *face—file—halt—front*, and take their place in the new column. In this manner, one or more battalions will have arrived, and be ready to form in the alignement.

But in order also to bring up the rear battalions, which are still in the old direction of the march—The filing of those platoons

is

is a signal for the ordering such rear battalions, each to *break* in column from the general one, by wheeling its head platoon, and directing its march upon its nearest flank point in the alignement, which must be previously ascertained by the detached adjutants. When there arrived such head platoon *wheels* into the new column, and halts; and the rear platoons of that battalion then *file* as above, and arrange themselves behind it. Thus each battalion arrives successively in a separate column in the alignement, and when so placed, pivots dressed, distances adjusted, the line will be formed by the platoons wheeling up the quarter circle.

Entry of the rear battalions of column into the alignement when the head halts in it.

Fig. 26.

---

In this manner the general column of march is resolved when necessary into the several columns of formation, and the leading platoons of each battalion are easily disengaged, whether they are marching at open, half, or quarter intervals, and directed on their several points to front, rear, or flanks, as the line is to be formed.

Where several columns of march are to form into one general line—Each will arrive at its ground in the same manner, but one is particularly named from which the rest take their directions and distance.

Arrival of several columns of march on the same alignement.

Fig. 31.
Fig. 32.

---

From the most exact attention and application of the rules prescribed did the SILESIAN infantry in 1785, consisting of twenty-three thousand men, enter the left of the alignement; march along it in column of half companies for about an hour and a quarter; halt at the signal of a cannon, and form in a most correct line, extending two English miles and a half. This line required no after dressing or adjustment of distances; but immediately marched in front about three hundred yards, halted, and fired, each battalion by grand divisions. On many other occasions in 1785 were lines of from five thousand to ten thousand men formed, in this manner, with the most perfect accuracy.

March and formation of Prussian infantry.

Fig. 67.

---

CHANGES

( 96 )

## Changes of Position of the Line.

All changes of Position of the line (not made by the march in front) from one diftant fituation to another, are effected, by the application, and movement of one, or more open columns, and may be divided into four parts.

<small>General principles of changes of pofition.</small>

1ft. The line wheels the quarter circle by divifions to either hand, fo as to be ready to break into one or more columns.

<small>Fig. 30.</small>

2d. The column, or columns file by divifions, or march in front as is neceffary and ordered, to arrive in the new direction.

3d. The divifions again form in general column perpendicular to the new direction.

4th. When the divifions of each battalion are arrived at their ground, halted, and adjufted—the line is formed by their wheeling up, and thus battalion after battalion.

---

When all the divifions of a battalion, or of a more confiderable column, *file* to arrive in the new direction—it is by the juft management of the two leading divifions, that fuch alteration of pofition is truly made. The principal attentions are,

1ft. That at all times the leading divifions arrive firft in the new direction, and there front perpendicular to it.

2d. That at every inftant of the tranfition from column to line, the wheeling diftances betwixt the divifions fhould be nearly preferved.

<small>General rules in the change of pofition of the battalion column by files.</small>

3d. That during the movement the leader of the fecond divifion *files* fteadily at the ordered pace on the given point of march C, in the new direction—The leader of the firft divifion, to whom the point of formation is afcertained (which may be in, but not advanced before the intended line) keeps that point, and the head of the fecond divifion, invariably in a line during the whole tranfition, in the fame manner that the two adjutants arrive at their intermediate

<small>Fig. 30.</small>

point

point in the alignement. The leading flanks of the rear divisions, never overpass those of the two front divisions, but endeavour without hurrying, to line with them exactly, and to arrive successively in the new direction, with their just distances.

By these means the commander who is himself with, and conducts the two leading divisions, places them in the direction that best answers his views, and at once takes up any position and to any front that is necessary. As circumstances change his intentions, he may at every instant vary, and direct them upon new points of march and formation; the rear of the column always conforming (without the necessity of sending particular orders) to whatever alterations of direction the head may take; and the commander conducting that head, so as to enable the rear to comply with its movements, without confusion or too much hurry. *Attentions of the commander.*

---

When one or two battalions only, are to take up a new position to front or rear: it will in general be best done by divisions wheeling to a flank and then all filing.

But where a line of several battalions is to make such change—it may break into columns of platoons, each of half, or whole battalions, and the heads of those columns will be conducted by a mounted officer, and by their separate routes to the points in the new position, where they are to extend into line. When the several columns are at any time marching in concert to gain a new position; their heads will be regulated and directed by the commander of the whole, in the same manner, and upon the same principles that the leading flanks of divisions are when filing into the new direction, as already prescribed. *General mode of position of a line.*<br>*Fig. 32.*

( 98. )

The OPEN COLUMN arrives at all times, and halts one flank of its divisions in the line of formation, in one of the four following situations.

**General situation of the open column, when prepared to form in line.**

**Fig. 39.**

Viz. with $\begin{cases} \text{The right in front, arriving} \begin{cases} \text{Before} \\ \text{Behind} \end{cases} \\ \text{The left in front, arriving} \begin{cases} \text{Before} \\ \text{Behind} \end{cases} \end{cases}$ The new lines of formation.

The term of arriving before, or behind the new line of formation is understood that—In the first case, the pivot flanks of the divisions of the new column are placed on the farthest side of that line.—In the second case, they are placed on the nearest side of that line. In both cases by wheeling to the pivot flanks they produce the line required, but to different fronts.

---

**Relative positions.**

All new POSITIONS of a line with respect to the old one, will be executed by the movements of the open column, and taken up from one of the above situations.

Such positions are $\begin{cases} \textit{Parallel} & \text{- or nearly so to the old line.} \\ \textit{Interfecting} & \text{- some part of the old line, or its prolongation.} \end{cases}$

The

( 99 )

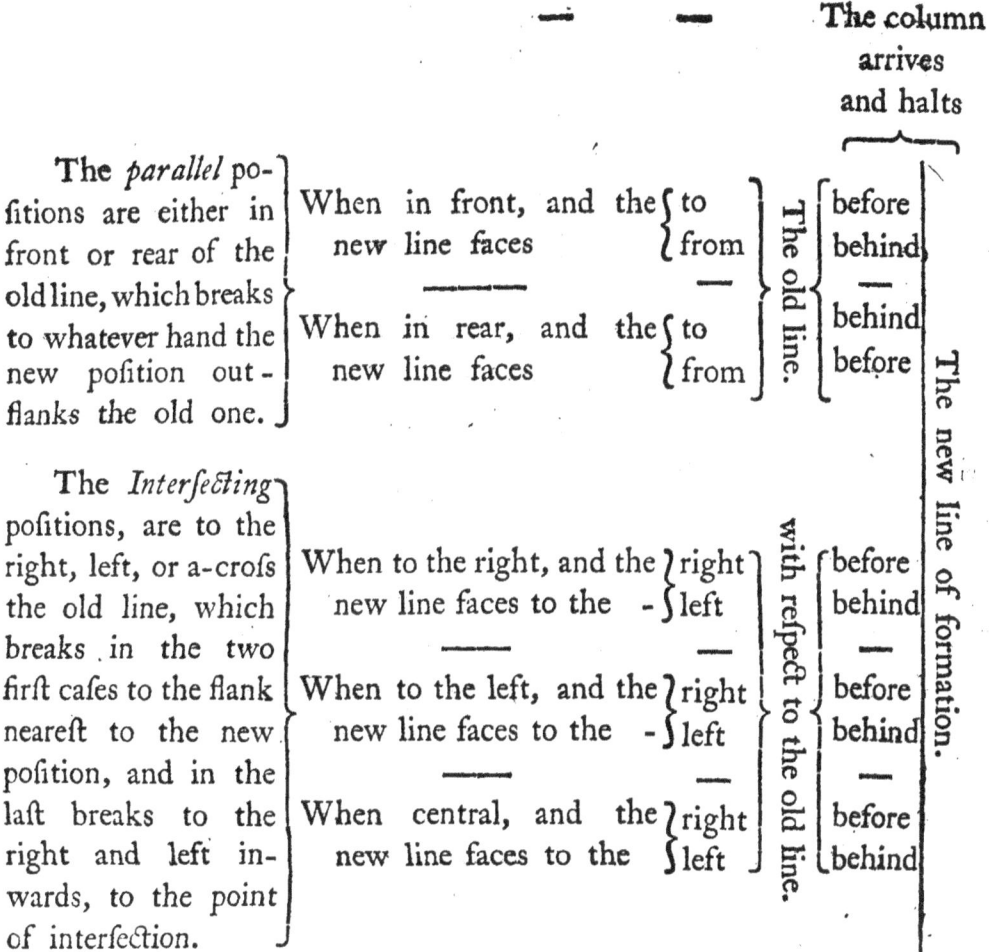

Fig. 40.

The above new positions are supposed parallel, or perpendicular or nearly so to the old one. But all the different intermediate degrees of inclination may be taken and resolved into the one or other class.

As on all occasions the line breaks into open column to whichever hand, and to whatever point it is to manœuvre: so if the battalion when formed is to take a new position, on one of its central divisions, or the line on a division of one of its central battalions; it will break to right and left inwards, and face such division.

*Battalion or line break to to the directing divisions when to change position.*

In central changes of a line, or central changes of a single battalion; such central or single battalion would appear to manœuvre easiest by the whole of it breaking and fronting to one and the same flank. But it must be recollected, that this would become an exception

*Reason of breaking to the directing division.*

O 2

exception from the general method of the line, and would not apply where the division of formation was moveable.

---

<sub>When from line a new position is to be taken on any one given central division a.</sub>

<sub>Fig. 33.</sub>

<sub>Fig. 34.</sub>

The named division is wheeled as much as is necessary to place it with its pivot flank perpendicular to, and on the new line of direction.
{
  If to front to the right; its right flank will be in that line, and it will face to the intended left.
  If to front to the left; its left flank will be in that line, and it will face to the intended right.
}

The line then breaks inwards to the named division a.
{
  The two divisions, viz. One on each side the named one, will only wheel one eighth of the circle, in order to avoid interfering, and the more readily to disengage their flanks.
}

The remaining divisions of the central battalion *file* into the new direction, and place themselves in front, and rear of the standing division. The other battalions *march* separately in columns of platoons from the leading flanks, and take their places in the general column. When standing in this situation, the right half of the line will always have the right flanks in the new direction, and the left half of the line their left flanks.

---

<sub>When with respect to the old line</sub>

<sub>The new line fronts to the left.</sub>

<sub>Fig. 35.</sub>

<sub>The new line fronts to the right.</sub>

The right divisions of the central battalion *file* to their right, and the right battalions each in column *march* to their right. They all place their right flanks in the new line, in front of, and facing to the named division a. The left divisions of the central battalion *file* to their right also; and the leading divisions of the left battalions *march* to their right. They all place their left flanks in the new line, fronting the same way as, and in rear of the named division a.

The right divisions and battalions move to their left, and place their right flanks in the new line, and in rear of the named division a. The left divisions and battalions move to their left also, and place their right flanks in the new line in front of, and facing to the named division.

The

( 101 )

The new line is formed by wheeling inwards. Each division to its flank placed on the new line.

As the named division, and whatever other one is next, and facing it in the column, wheel inwards to form the line; such division in taking its place in the column, must be careful to allow for this circumstance, and have an interval equal to twice its front.

Fig. 35.

---

The divisions which are then in front of the named divisions, will *countermarch* and face it—That division will then be placed, and the new position taken up, as above prescribed in changes of the line.

When from open column the line is to be formed on any central division and to any one point.

---

If in the course of taking a new position on a central division a, it is meant that such divisions should not remain fixed, but be MOVEABLE either to front or rear, with respect to the old line.

The named division a, will be first wheeled and placed with its pivot flank, perpendicular to, and on the new direction, and fronting the way the line is to extend. If to the rear, it must countermarch, and face that way.

The line will then break inwards. {The divisions on each side the named one, will only wheel the one eighth of the circle.

When from line, a new position is to be taken on any one given central division, and which division during the operation is moveable to front or rear, in order to extend the line to a particular flank.

The named division a, will now be put in march along the new direction, and be followed in double (though perhaps unequal) column by the remaining divisions of the central battalion, and covered by one of those columns, viz.

Fig. 36.

If

If marching to the front, and
- To face to the right { The right column of divisions of the central battalion will cover it; and the left column will march along side of it, to its left.
- To face to the left { The left column of divisions will cover it, and the right column will be on its right.

If marching to the rear, and
- To face to the former right { The left column of divisions will cover it.
- To face to the former left { The right column of divisions will cover it.

Fig. 36.

When it arrives at its point d, it will halt—Those divisions which are to be in front of it (and are then marching by its side in column) and now separated by a distance of three or four paces, will move on, and place themselves in column before and facing it—Those which ought to be in rear of it, will already be in that position. The other battalions which moved when the central one did, and which in the mean time have been marching in separate columns, by their leading flanks, and pointing to front or rear relatively to the central battalion, approaching but not entering into its direction; those battalions will then take their places in the general column, in the new line as already prescribed.

Fig. 37.

If the named division is the flank one of a battalion, in that case, the whole central battalion will follow it in one column only, and the adjoining battalion will compose another column, and march a-breast of it, separated by three or four paces, till the named division comes to its ground and *halts*—The adjoining battalion will then proceed and take its place in the general column, in order to form into line.

Indeed, in all cases where the named division is to be moveable, the whole of the central battalion, may without much loss of time, be thrown into column before and behind that division; and the divisions in front of the named one *countermarched*, in order that

that the whole may face the way the column is to move—This done, the general movements of the central and other battalions, separately by column may begin; and in such case, the front division of the central battalion, will be the first to arrive and halt at the point where the column is to extend into line.

*The central battalion of a line may always lead from a flank in the change of position.*

This movement of the given division, is equivalent to the line marching from the center, either to front or rear, and from that situation forming away to the flanks—It is the kind of movement, which a second line must often make, in order to comply with a change of position in the first; or, to the line first marching forward or backward, and then making a central change on a fixed point.

*Fig. 38.*

---

From what has been mentioned, the following *general method* should take place in all changes of position of a considerable line.

The line breaks into column to whichever hand the new position outflanks the old one—The several columns of battalions are then put in march—The leader of the second battalion from the directing flank, has a point in the new line ascertained to him, and marches upon it. The leader of the flank battalion will preserve the parallelism, or give gradually the new inclination to the heads of the other battalions—These nearly dressing up, and preserving their battalion distances, arrive, and throw themselves again into general column in the new direction, and then wheel into line.

*When the new position is in front or rear of the old one, and parallel or nearly so to it.*

*Fig. 40.*

If the new line is to be in front of, and facing towards the old line, the battalions within themselves will not be inverted, but the right of the line will now be the left; nor can it be well avoided, unless by countermarching the line before the movement, or, by a complicated operation during the movement, or, by countermarching first the battalions, and then the line after the movement.

The

When the new position or prolongation of it, intersects to the right or left of the line.

Fig. 40.

The line will break to whichever flank is nearest to the new position—The heads of battalion columns will be separately conducted to their points in the new line, being regulated by the leading flank battalions—They will again enter into the general column, and form in line by wheeling up—or, the line after breaking to the flank, may continue its march in column to the point of intersection, there wheel into it; and when the head halts, the rear battalions will separate from it to take their places in the new line.

When the new position or prolongation of it, intersects the body of the line.

Fig. 40.

The division which is in the point of intersection, will be moved out and placed perpendicular to the new direction—The line will break inwards towards that division—The divisions of the central battalion will file, and place themselves in column before and behind the standing divisions—The other battalions will be conducted separately in column to their point in the new line; there throw themselves into general column, and form into line when ordered.

If it is meant during the movement to extend the new line towards either flank, it will be done as before prescribed.

---

General method of forming a considerable open column from line on any division.

Fig. 41.

When from line, an open or half open column of several battalions is to be formed on any given division, and in any direction—After breaking inwards, the divisions of the battalion of direction will *file* into the new column. But the other battalions will *march* diagonally and separately towards their several points; and when the head of each arrives in the column, the rear will file into it. The front divisions of the column will countermarch, and the column will then be ready to move on.

---

General attentions in the open column and in change of position.

It cannot be too often repeated—that the preservation of wheeling distances; the covering of the pivot flanks when marching in column, and particularly when halted to form; and the just wheels of the quarter circle from line to form in column, or from column to form in line: are the great and essential circumstances that determine exactness in all movements of the open column, or in all changes of situation in the line from one position to another—They cannot be too scrupulously adhered to, all just and quick formations depending upon them, and all after dressing and unsteady shifting, being thereby avoided.

---

When the GROUND is perfectly open, and such as will allow of the movement of the line, or its parts in front; then a change of position may be made on a flank, or on a central point by the march of divisions in front—In such case, some one given division is accurately placed in the proposed direction, and the other divisions of the line *wheel* separately an equal portion of the circle so as to be each parallel to the given division. Those that are behind it *march* up in front, and line anew with it; those that are before it, *face* to the rear, *march*, line with the given division, and then *front*—All the divisions move at the same pace, and at any instant ordered, *halt, front, wheel*, and are formed in line.

<small>Change of position in open ground by the wheel up and march of divisions in front.</small>

<small>Fig. 42.</small>

This mode of changing position, is more immediately applicable to the movement of a battalion or of a small body, where the point of forming is fixed, the change of direction inconsiderable, and the ground perfectly unembarrassed—But the change of position of greater bodies, will be commonly effected by the more general method of their breaking into several columns, marching to their several points, and then extending into line. In some situations from open column, the divisions may thus by oblique marching, move up into line.

In this manner, at *Berlin* and *Magdeburg*, two corps of six and of thirteen battalions, changed their front on a central division; and in the course of the operation being attacked by cavalry, they several times, fronted, formed into line, fired, and again resumed their march.

The MOVEMENTS of the open column, are particularly neceſſary on the following occaſions.

When a line formed in order of battle, is to extend in the ſame direction to either flank, in order to follow the movements of an enemy, or to outflank him if he remains poſted.

*Situations which excluſively require movements in open column.*

When arriving in column of march on any ground, the commander in determining the general direction that his line is to take, ſhall not have been able to aſcertain the points where he would fix the flanks of it; but after entering into it, is obliged in conſequence of the poſition or manœuvres of the enemy, either to ſtop his own movement ſooner than he intended, or to prolong it beyond the point he originally meant.

*Fig. 27.*

When after ordering the head of the column to halt and form in line, and that the rear battalions ſhall have broke from the column, and are marching into that line; he ſhall find it neceſſary to reſume the column of march in the former or any new direction—Then ſuch part as may be formed in line will again break, and thoſe rear battalions will relatively alter the direction of their heads, and take their places as before in the continuation of the column; and any diviſions that are ready to wheel up and form, will march by their flanks, and throw themſelves into their proper places in the column of their battalion.

---

*Advantages when diviſions wheel backward from line into column.*

Although the GENERAL rule and method for wheeling from the line, when halted into column, is for the flanks of diviſions, which are to become the pivot flanks of the column, to wheel forward to the hand to which the line is to break; yet there are occaſions, where it is advantageous for that flank to remain fixed, and for the other flank to wheel backward, and thereby acquire the intended poſition in column—Small diviſions will perform this by wheeling back, but ſtill preſerving their original front—Larger diviſions muſt *face* to the rear, *wheel* to the pivot hand, *halt*, and *front*.

*Fig. 28.*

By

By this means, although the divisions should be unequal, the pivot flanks will remain exactly dressed in the direction of the former line, and will be in readiness to march along it. In this manner, a battalion formed in line, will throw itself into column when ordered to extend to either flank, along the prolongation of that line—In this manner also, the divisions forming the flank sides of the square or oblong, will wheel back into column, when the body is to be put in march.

It is therefore essential, that the battalion should be practised and ready at this movement, which on many occasions is necessary—And so important is the circumstance of having the pivot flanks always dressed, that was it not in compliance with established custom, this method would be proposed as the general one of forming column to the flank from the line, when halted.

---

It would perhaps answer every purpose of wheeling in a COLUMN in MARCH, and prevent the extension of it, to order the leading division to wheel at a small increased pace, and proceed—Each succeeding one, will in such case, continue to look to, and wheel by the pivot flank; which pivot flank, whether the wheels are made to right or left, will only have to observe this one general rule; never to allow the preceding division to increase its distance from it, and will therefore on the wheel move quicker or slower, consistent with that attention.

*A method for divisions wheeling in the column of march.*

When the pivot flank is not the wheeling one, it will easily follow the movements of the preceding division in the circular direction.

When the pivot flank P is the wheeling one, the other may still be in some degree moveable, and describe during the wheel, a quarter circle on a given point C, three or four yards beyond its flank, by which means each division will be free of, and not interrupt the movements of the succeding one.

*Fig. 29.*

The words *wheel*, and *forward* when that wheel is compleated, will be given to each division.

The wheels in column at half and quarter diſtance, can beſt be accompliſhed by this general attention, in the following diviſions, not to allow the preceding ones to increaſe their juſt diſtances.

# OF THE
# CLOSE COLUMN.

THE great object of the CLOSE COLUMN, is, to form the line to the front in the quickest manner.

The close column, is generally formed from open column of march or manœuvre, when the head of such column is arrived at, or near the ground where the line is to be extended. Fig. 43.

In this shape, the several bodies of the army assemble in presence of an enemy, at a sufficient distance, and avoiding the enfilade of cannon—Their numbers cannot then be easily discovered— They are under the eye, as it were in the hand, and ready to execute the final arrangements of the commander, after he has examined the appearance of his opponent—With the greatest facility they take up any position, and the assembly and formation of infantry columns, are always protected by cannon, cavalry, and such advantages as the ground offers. Chief objects of the close column.<br>Fig. 44.

It can never be necessary or proper, to march any considerable distance in close column—But when the formation of the line is to be to the front, and where there is nothing to apprehend for the flanks, the column at half or quarter intervals may generally be that of march, and the close column will be formed, when its head arrives at the ground on which it is to deploy into line— March of column.

The

( 110 )

The clofe column may occafionally be required to march fome fmall diftance to its flank; to correct intervals; to gain an enemy's flank; or for fome other particular purpofe. But a confiderable movement in front, can hardly be expected from it without loofening its divifions.

Front.

*Deploying* from clofe column, though in many fituations, a moft accurate method of taking up a given and fixed pofition in line, is not on all occafions to be preferred to the marches of the open column; for if once the operation is begun, the line muft be compleatly formed before any other movement can commence, and which a change of circumftances may fometimes require.

When confiderable bodies are to form, the clofe column fhould not be compofed of lefs than two companies in front; but in the movements of fmall bodies, it may be compofed of companies or platoons.

Diftances.

When the column is halted—Rear ranks are one foot, and divifions one pace afunder—Mufic, drummers, pioneers, &c. are clofed up to the feveral divifions they cover when in line—Officers, fergeants, &c. are in their places as in open column—Staff officers are on the flanks of the column. In this fituation, the feveral divifions are ready to *face* outwards, and *deploy* into line.

There will be a diftance of two paces from battalion to battalion when halted in clofe column to *deploy*.

Commands.

The clofe column will receive and execute all orders directly from the commanders of battalions; nor will they be repeated by leaders of divifions, unlefs abfolutely neceffary.

All filings from line to form in column, or from column to form in line, muft be made at the quick ftep.

The clofe column fhould not if poffible exceed five or fix battalions; when there are more troops, better to form more columns

if

if it can be done; and therefore the columns of march may often be subdivided when they come near the points of forming into line, directed upon them, and then closed up. — Strength.

If there is artillery marching with the column, it must take up as small a space as possible; and in general it may be placed on its flanks till ordered to its position. — Artillery.

The column will dress and cover to the left, if the right is in front; and to the right, if the left is in front, following the same rules as the open column.

The column at half or quarter distance must always preserve an interval betwixt battalions, equal to the front the column is marching at. — Intervals.

## Formations of the Close Column.

The Close Column is formed either from line, or from open column.

---

When the close column is formed from line—any one division is named, as that on which the column is to form, and which division may be placed obliquely if necessary. — Close column formed from line.

A caution will be given, whether the right or left is to be in front. The whole except the named division, will then be ordered to *face* inwards, and the three leading files of each division will throw themselves (according as they are to be in front or rear of the standing one) to right or left, till disengaged from the ranks of the division preceding them. — Fig. 45.

In this situation, being enabled to move independent, each will take its point, and be ordered to *march—close in—halt—front—dress,*

*From line to form clofe column.*

*drefs*, and place itfelf properly, before or behind the ftanding divifion, fo that the pivot flanks fhall always cover; and thofe that are in front of the ftanding divifion, muft take their diftances from the rear.

Where feveral columns are in this manner to be formed by parts of the fame line—fuch parts and the ftanding divifion of each will be fpecified. The fame flanks will be in front, and the columns will form feparate and independent.

Fig. 41.

---

Fig. 43.

When from open column, the CLOSE column is formed, in order to deploy into line, the following fucceffive operations take place.

1ft. When approaching the alignement, the platoons are ordered to *clofe* up to quarter diftance to the leading one, which is either halted when neceffary, or only directed to fhorten its ftep.

2d. The column then *moves* on, till its head *halts* exactly in the alignement, and its rear *marches* up clofe.

*From open to form clofe column.*

3d. Every other platoon (from the head of the battalion) is ordered to *face*, and *double* up to right, or left of the one preceding it, according to the formation of the column. The commander of the right platoon takes poft on the right of the divifion. The commander of the left platoon on the left divifion, and a covering under officer in the center of it, there confequently is an officer on each flank of each divifion.

4th. The whole again are ordered to *clofe* up to the front, and in this fituation the column remains ready to extend into line.

---

*From clofe column to form battalion.*

From CLOSE COLUMN the BATTALION is formed by three manners of *Deployment*.

Viz. on $\left\{\begin{array}{l}\text{The front}\\\text{The rear}\\\text{A central}\end{array}\right\}$ Divifion of the column.

In

In each cafe, the named divifion is that of *appui* from which the reft extend, and to which they form and drefs.

The mode of formation remains the fame, whether the column has its right or left in front; allowance only being made for the inverfion of circumftances, and the neceffary changes in the words of command.

The battalion is here fuppofed to have its right in front; to have performed the feveral operations already prefcribed, neceffary to its being placed on the alignement ready to form; and that the clofe column confifts of eight divifions.

---

When to form on the FRONT divifion—The column being halted with its front divifion in the alignement, receives the caution that the *line* will be formed on the front divifion—That divifion remains firm, and the others are then ordered to *face* to the left, and to *march* with their leading files dreffed and parallel (not oblique) to the line of formation—When the left leader of the fecond divifion has marched a fpace equal to the front of his divifion; he orders it to *halt—front—eyes* to the right—The officer on the right then gives the commands: *march—halt*, when arrived on the alignement, and that he joins the left flank of the right divifion—He then as already prefcribed, dreffes his divifion by the ftanding one, and the diftant given point to the left. In this manner, every other divifion proceeds, and each fucceffively *fronts* when it has marched fufficiently to the flank—*moves* up—*halts* and *dreffes* in line. *[On the front divifion. Fig. 46.]*

---

When to form on the REAR divifion—A caution is given, that the *line* will be formed on the rear divifion. Two non-commiffioned officers are immediately fent from that divifion, to place themfelves correctly clofe before each flank file of the front divifion; and the leader of that divifion, is fhewn the diftant points in the alignement, on which he is to march. The word is then given, to the right *face*; on which all the divifions, except the rear one, face to the right—At the word *march*, the faced divifions ftep off, the *[On the rear divifions. Fig 47.]*

heads of their files dreſſed, the front one moving in the alignement, and the others cloſe to its right.

When the rear diviſion is uncovered, it receives the word *march*, and then moves up to the two under officers, who mark its new poſition; it there *halts*, and is dreſſed very exactly in the line, its left flank being the point of appui.

When the leader of the ſeventh diviſion, which preceded the rear one, has marched to the flank, a ſpace equal to the front of his diviſion, he orders it to *halt—front—eyes* to the left; the officer on the left, then gives the word *march—halt*, when arrived on the alignement, and that he joins the right flank of the left diviſion, from which he dreſſes his own in the alignement.

All the others ſucceſſively proceed in the ſame manner, till the right one, which has been marching on the alignement, *halts—fronts*, and *dreſſes* in it.

---

When to form on a CENTRAL diviſion—The commander having determined according to circumſtances, the number of diviſions he means to throw to right or left, *names* any one given diviſion of the column, as that of formation. For example the fifth. It immediately detaches two under officers to mark its ground in the front, and the leader of the head diviſion, takes his point to the right in the alignement.

On a central diviſion.

The command is then given—Diviſions outwards *face*; on which, the four front ones face to the right, the three rear ones to the left, the central one remains as it was. At the word *march*, the faced diviſions move in file to the flanks; and when they have uncovered the central one, it moves up and *halts* on the alignement as before directed—The front diviſions proceed as when the line is formed on the rear of the column, and the rear diviſions proceed as when the line is formed on the front of the column.

Fig. 48.

---

The whole column *files* to the flanks, at the word from commanders of battalions; and it will require great attention in thoſe of diviſions to front, cloſe in if neceſſary,

and

and drefs fucceffively when oppofite their proper ground in line, always taking care not to overpafs whatever divifion precedes them in the formation.

The leader of the front divifion muft file critically along the alignment, the other front divifions move parallel to, and behind him; and the heads of the whole continue dreffed till each fucceffively halts, and moves up to its place in line—The fame directions alfo regard the rear divifions, which are regulated by their leading one.

In filing, the divifions muft not open out, but march quick, compact, and at an uniform pace—The juftnefs of formation depends on officers timing their commands, judging their ground, and on their particular attention.

*Attentions in forming the battalion.*

The major or adjutant mounted, accompanies the deployment to affift in fronting the divifions, in cafe the leader of each is not critical in his commands, or that he is not heard, or that his files are too open; and the defects alfo of a preceding divifion may be remedied by the judicious ftop of the one following it.

As the commanders of the platoons are on each flank of the divifion: to whichever flank the divifions file, the outward officer will halt and front them; and the inward officer will move up, halt, and drefs them in the line; they will then refume their places on the right of their platoons.

The line into which the clofe column extends, muft be always *perpendicular* to the ground on which the column ftands—When that line is afcertained, the column muft be placed accordingly before the formation begins, and the neceffary points in it muft be previoufly fixed.

When a Column of March of several battalions (the right in front) is to form in close column, and from thence extend into line, the following operations take place.

Fig. 49.

1st. The column being here supposed of five battalions, and marching by platoons, is ordered to close to the front to quarter distance.

2d. As the column will generally arrive at an intermediate point of the ground where it is to form, one of the battalions (as here the third) is named as the central one, on which the column is to deploy. When therefore, the head of it is arrived within two hundred paces of the point where it is to be placed in the alignement, the whole *halt*—The same operation is then ordered for the column, as takes place when the single battalion deploys on a central division into line, viz. The two front battalions *face* and *march* to the right—The two rear battalions in the same manner to the left—The central battalion when uncovered, *advances, halts* its head in the alignement; and the other battalions allowing it to arrive there first, successively, *front, move* up, and *halt* also in the alignement, with an interval betwixt each, equal to the front of two platoons, which they acquire during the march to the flank, and before they front to move up. Artillery, music, drummers, &c. are in the rear of each battalion column, and their several heads are accurately dressed in the line.

From column of march to form close column and the line.

3d. Every other platoon is ordered to *face, double* to the left of the one preceding it, *form* divisions, and the whole to *close* to the head of their respective battalions.

4th. When in this situation, any one division of the central, or of any other battalion is *named*, as that on which the line is to be formed; here it is supposed the fifth division of the third battalion. At the word outwards *face*, the two right battalions, and the four front divisions of the third battalion, face to the right; and the two left battalions, and three rear divisions of the third battalion, face to the left, the fifth division remains fast—At the word *march*, the whole move on in file, and the third battalion

forms

forms in line, as directed on the central division. The second battalion, when it has marched a space equal to its proper interval in line, beyond where the right of the third battalion is to be placed, *halts—fronts,* and being then on the left of its ground *deploys* into line on its own rear division; and in the same manner does the first battalion *deploy,* when it has halted at the left of its ground. The fourth battalion, when it has marched a space equal to its proper interval in line, beyond where the left of the third battalion is to be placed, *halts—fronts,* and being then on the right of its ground, *deploys* into line on its own front division; and in the same manner does the fifth battalion *deploy,* when it has halted at the right of its ground.

From column of march to form close column and the line.

The points of the alignement must be exactly ascertained in these operations by the adjutants, and the majors will much assist in giving the just distances to the battalions before they deploy. Those divisions that march on the alignement, must be very careful not to get before it. As the distance of a platoon remains betwixt each battalion after the divisions are formed, it allows for any loosening of the files that may take place in a considerable march to the flank—The officers that conduct the divisions next to the front, will step out successively from their divisions when uncovered, and thereby see more exactly the proper instant of ordering their divisions to *halt* and *front.*

Attentions in forming the line.

In a considerable column, the deployment will generally be calculated to take place on the front, instead of a central division of the named battalion.

---

Should a considerable close column be halted short of where it is to form into line—The several battalions may be first disengaged from each other, and placed in the *echellon* position; they may then be marched separately to their several flank points in the alignement, *halt,* and *deploy* to right or left to occupy their place in line.

Forming the line from close column.<br>Fig. 50.

When

**Formation of the line from two columns.**

When Two columns marched off from the right of wings are required to form in line to the front, flank, or rear of their march.

**To the front.**

**Fig. 51.**

If to the FRONT—Points in the alignement and relative to the flank point of *appui* are given by adjutants, at which the head of each column is to be placed. The distance betwixt those points being known to the chief commander, as soon as the battalions of each column are placed in the manner already prescribed on the alignement, he *names* the division of a certain battalion of each column on which the rest are to deploy. The operation as already directed takes place; the allotted portions of each column *deploy* inwards, and fill up the vacant space; the rest deploy outwards, and the line is compleated.

**To the flanks.**

**Fig. 51.**

If to either FLANK—the points in the alignement at which the heads of the columns are to be placed, are immediately ascertained by adjutants—The columns are closed up to quarter distance—Their heads are wheeled when proper, and directed so as to arrive and *halt* just behind the alignement, with the whole of each column perpendicular to it—The battalions are then drawn out in *echellon* on the named one—*move* up to the alignement, and *deploy* as already directed.

**To the rear.**

**Fig. 51.**

If to the REAR—The columns by such operations as are necessary are closed to quarter distance on such division of the column as is near to, but before the line of formation. The whole or such parts of the columns as have not countermarched in order to close their distance, will then *countermarch* by divisions, and the columns will stand as marched off from the left and fronting to their former rear—The columns will then be ordered to *exchange* places, by the divisions of each *facing* inwards and *filing*; when they have nearly approached each other, the right column *halts*, and the left one continues in march, and passes through the intervals of the divisions of the right one; both columns then *move* on, until they

*arrive,*

*arrive*, *halt* and *front* on the ground which each other occupied, and which has been properly marked and preserved for them: during this flank march, the heads of the files are kept nearly dressed, and are regulated by the two divisions of the head—The columns are now in the situation of having marched off from the left of wings, and of forming in line to the front, they will therefore proceed as before directed.

This method of exchanging columns by files may be applied to advantage, when in the course of a night march, one column may have crossed the line of another, thereby lost its relative situation, and that it is necessary to regain it.

---

If there are Two columns composed each of part of two lines—the battalions of the second line will halt at a proper distance from the first, and deploy in the same manner.

<small>Formation in two lines.</small>

<small>Fig. 52.</small>

---

Where more than two columns are to form in the same line——when halted, their heads must be at considerable distances. If the distances of the heads contain more ground than the interior columns can occupy, then part of the exterior ones will of course deploy inwards. If the interior columns are more than sufficient to occupy such space; part of those columns, and the whole of the exterior columns will *march* to the flanks, *halt* when proper, and then *deploy* into line.

<small>Formation in one line from several columns.</small>

<small>Fig. 54.</small>

The same general directions regard the heads of open columns, when several of them arrive at different points in the alignement.

---

In *deploying* from close column, its head must ever first be halted on the proposed line; and the Division on which the formation is made, must, during the operation of forming, always *advance* to the point where the head of the column was so placed. The mode of extending into line from close column by making the front, divi-

<small>Fig. 56.</small>

sions

**Guibert's formation of the line from column defective.**

sions *file* towards the rear, in order to dress with that of formation (which remains fixed, and immoveable) is defective, false, and cannot be adopted as a general method, although prescribed by so respectable an authority as that of GUIBERT.——Because——

**Fig. 55.**

It is an unnecessary and improper operation, first to bring the column forward to the intended line—then to send the head of it back in order to form—and again to advance the line to its original point, where it ought at once to have extended—The difficulty of the divisions getting into line, by these retrograde motions is increased, as the directions are less certain, and the flank points not sufficiently determined and known; for at no time is the front of a single division a sufficient direction for forming of any one battalion.

**Reasons.**

**Fig. 57.**

The true line is that to which the commander, after having well reconnoitred, and determined the position he means to take, has brought up, and pointed out to the head of the column; and which during the unfolding of it, is preserved for the division of formation, by the two non-commissioned officers who have been dispatched from it to the front, before the movement begins—If such division was not to move up, but remain *fixed*, the new line would probably not even be parallel to the true one, and subsequent operations would be required to give it the intended position.

But above all the impropriety of this method is apparent, when *several* columns are to extend in the same line—For the heads of them being halted in the new line; circumstances will determine the commander, according to the number of troops necessary to fill up the intervals, or the numbers he means to throw to the right, or left, to order on which division of each the different columns are to *form*—It is therefore evident, that should such divisions remain *immoveable*, and not march up to the front, the several columns when *deployed* will be formed in as many separate lines; those probably not even parallel among themselves, and all *retired* behind the intended line.

The PRUSSIANS were formerly accustomed to begin the *deploy* from the close column of several battalions in depth: but the previous operation of drawing out the battalions, and placing each separate on the alignement, now takes place; and when each battalion has arrived at the point where it is to deploy into line, it is only four divisions in depth. The *deployment* of a considerable column is more practised as a cavalry than as an infantry method of formation; the battalions of infantry most commonly break from the general column, march to their several fixed points in the alignement, enter into it, take the wheeling distances of divisions, and then form, as prescribed for the open column.

Prussian mode of deployment.

---

At MAGDEBURG in 1785—A corps of infantry marched off, in two columns from the right of wings, each column composed of half of each line, and each line of ten battalions; a vanguard of two battalions preceded the left column, and marched in full front: the cavalry in column covered the front, and the right flank of the right column. The columns were one eighth of the battalion in front, and closed during the march to half distance.

The columns halted about 1700 yards from where the opposing corps was posted—The divisions moved up to close column, and the leading battalions of the second line preserved their relative distance from the front of the whole. The battalions then drew out from the columns on the center one of each, halted, faced, moved up to the alignement, and halted in line of battalions; those of each column at the distance of a division from each other, and an interval equal to the front of three battalions separated the columns. The second line performed the same parallel movements.

Prussian formation of two lines from column of march, and close column.<br>Fig. 68.

( 122 )

In this situation the whole moved on, obliqued to the right, preserved the general front, and after advancing about three hundred yards halted on the line prepared by the adjutants. The alternate divisions doubled up; one quarter of each battalion was now in front, the interval betwixt the columns remained, but the battalions of each column were without any interval, and the divisions closed up to the front.

The lines were ordered to *deploy*: the whole on the front divisions of the third and sixth battalions of each line (which stood fast) viz. The first and second battalions extended to the right, and formed on their rear divisions; all the other battalions of both columns extended to the left, and formed on their front divisions. The two lines of ten battalions each were formed with the greatest precision.

---

Deployment of the Prussian cavalry in two lines.

Fig. 70.

In SILESIA 1785—Thirty-five squadrons of cavalry, making about 4000 men, marched from camp in two columns formed from the center of wings, the front of each was composed of two half squadrons. They halted in the alignement at their proper distance, doubled the front of the columns by moving up every other half squadron, and closed up to the front. The columns being now each two squadrons in front, were ordered to *deploy* to right and left on their leading squadrons, which remained halted. In less than two minutes one full line of thirty-five squadron was most correctly formed, extended above an English mile, and was ready to move on to the attack.

Fig. 69.

Twenty squadrons of cuirassiers, and twenty squadrons of hussars formed in two lines, marched off in column from the center of each line. The column was two squadrons in front, the divisions closed up, the lines kept distance, and the head halted at its point.

The

The column *deployed* equally to right and left, and in ninety seconds both lines were completely formed *en muraille*, or without intervals—The cuirassiers in the first, the hussars in the second line.

---

The SILESIAN infantry of thirty-two battalions encamped in two equal lines, marched from the right to the front in four columns each composed of equal parts of both lines. The heads of the column halted in a given direction, at considerable given distances from each other, and the rear of each moved on to close column, the leading battalions of the second line preserving a sufficient distance.

The three front battalions of each column, and of each line faced to the right—marched to the flank—fronted, and moved up again into line, when separated from each other by a space nearly equal to the front of a battalion, the rear battalions when uncovered also moved forward to the alignment.

*Prussian formation in two lines, by movements of the close and open column.*

The whole battalions now stood in line in their regular order from left to right, and occupying an extent nearly equal to that of the line when drawn out.

*Fig. 71.*

An *oblique* position in front was now ordered for the formation of the line, by which the right was to be considerably thrown forward, and the point d'appui to be at the head of the left column.

The front division of the left battalion, first wheeled into the alignement, was followed by each other division, as it acquired its wheeling distance, and the battalion halted in open column when its left division arrived at the point d'appui. Each other battalion of the line continued to march forward, successively entered the alignement at the point where its left was to be placed, was followed by its proper divisions, and halted on its ground. In this manner both lines entered, and stood on the alignement in open column, when on the signal of a cannon the whole wheeled by divisions to the left, and formed in the most exact manner.

Such was a manœuvre of the Silesian infantry encamped in 1785. The left of the line was here the point of appui—the general distances were taken from the rear—and the quickness and exactness with which the adjutants lined in an instant from left to right, and gave the distances to their battalions was very conspicuous. The manœuvre was a mixture of the movements of the close and open column; and in this manner also at POTSDAM did a corps of cavalry and infantry halted in close columns behind the left of the alignement, form in two lines, extending to the right.

---

*Prussian march of four columns into the alignement and formation in two lines.*

*Fig. 72.*

The SILESIAN infantry marched in four columns from the right to the front, each composed of part of both lines. The leading division of the right column, wheeled to the right into the alignement, marched along it followed by the rest of the divisions and battalions. The other columns marched also to their determined points in the same alignement, and severally entered it on the left of their ground. The leading divisions of the second line wheeled when at their due distance, and marched on a parallel alignement to that of the first. The distance betwixt the heads of columns was such that the front of each joined the rear of that to its right when the whole had entered the alignement. The march was continued till the whole had entered, when at the signal of a cannon the whole halted, wheeled to the left by divisions, and formed in the most accurate manner in two lines.

The first line was perfectly formed in fourteen minutes from entering the alignement. The second line required twelve minutes more. The general front was about two thousand two hundred yards, and the whole body of infantry about twenty-two thousand men.

---

In this manner (as well as on other occasions) did the PRUSSIAN army in four columns covered by a strong vanguard, cavalry, and a cannonade, march on, and threaten the right of the AUSTRIANS at Liſſa. When at the diſtance of about a mile and a half, they ſuddenly wheeled the heads of the ſeveral columns into an oblique alignement, continued the march to the right by lines, gained the enemy's flank, formed with all the exactneſs of a review day, immediately moved on, and attacked the AUSTRIAN left, who in hurry and confuſion endeavoured but in vain to counteract ſuch regular and determined movements.

<small>Similar movements made at the battle of Liſſa.</small>

# MOVEMENTS

## OF THE

## BATTALION in FRONT.

THE MARCH of the battalion in FRONT is the moſt difficult, and the moſt material of all manœuvres; it is required near the enemy, and immediately precedes the attack.

The circumſtances that enſure the exact advance of the bat-talion, have been detailed in the article of march, muſt be ſcrupulouſly adhered to, and need not here be repeated. *March of the battalion.*

As the march of the battalion in front, is always parallel to the line on which it is ranged; it muſt be formed, halted, and moſt correctly dreſſed, ſquare to any particular point on which its future movement is to be directed: or, that point muſt be determined, in conſequence of the then line of the battalion.

---

A caution is given by the commanding officer, that the *battalion* will move *forward*. The firſt colour immediately advances three regular paces, and halts; the ſecond colour moves up into the front rank, and the two covering ſergeants paſs it from the rear, and place themſelves on each flank of the firſt colour. As under the direction of the commanding officer, the right advanced *When halted and to march in front.*

ſergeant

fergeant is to be anfwerable for the ftep, and for the prefervation of the line of march; the major, who is now at the center of the rear of the battalion, obferves and indicates the particular point, which is directly in front of that fergeant, and on which he is to march, after having remarked his intermediate points, which he may at all times find.

The colour and the fergeants, are placed perfectly fquare to the front on which they are to move, and parallel to the battalion. The commanding officer is advanced four paces before the colour, to have a general attention, and direct the movement of the battalion.

March in front.

At the word *march*, the whole battalion inftantly ftep off with the left feet, and turn their eyes to the colour in the front rank—That colour keeps its exact diftance of three paces from the advanced one, and is the guide of the battalion. The advanced fergeants and colour, give the utmoft care to preferve their parallel pofition, to maintain the equality of ftep, and to move on their given points; they are to allow no other confiderations to diftract their attention, and will notice the directions of the commander only, or thofe of the major in the rear, who is in a fituation to obferve and point out any defects that may arife in the march. Officers in the ranks, can only be obfervant of their own perfonal exactnefs of march; they muft certainly err, if they then too much attempt to correct inaccuracies in their divifions, that muft be done by thofe who are pofted in the rear; and well trained foldiers of themfelves know the remedy that is required, and will apply it.

The great points that infure the exactnefs of the march in front, cannot be too often repeated; the perfect fquarenefs of each individual's body; the touch of the files; the exact perpendiculars of march; the attention of the advanced fergeants and colours to the line of march, cadence of ftep, and parallel movement; and that of the battalion to cover, follow, and in every refpect comply with the motions of thofe that are advanced for their direction.

Inattention,

Inattention, or an inequality of step, will produce a waving in the march of the battalion; but the communication of this may often be stopped by the exertions of the major and adjutant, who seeing where it originates, will immediately apprize the platoons in fault, and coolly caution the others that are still in their true line, not to participate of the error. A flank of the battalion, may at first sight be thought to be behind, when the fault really arises from a center platoon bulging out, and thereby preventing the flank from being seen.

Attentions in the march.

---

The battalion thus marching in perfect order, when it arrives at its given point, receives the word: *battalion*; on which it ceases to gain ground, marks the time, and corrects any irregularity that may exist; after a pause of five or six seconds, it receives the word: *halt*. The step which is taken is instantly finished, eyes remain turned to the center, the sergeant and colours fall back to their station in the ranks, the whole remain steadied, and the commanding officer places himself close to the front rank, in order to see whether the battalion is sufficiently dressed, and in a direction perfectly parallel to that on which it marched.

When in march and to halt.

---

But if he finds it necessary to order a farther dressing, as the colour is a fixed point, which cannot be wrong, nor ought to be moved; he immediately himself takes three exact paces from the colour to the front, and from thence being uninterrupted by the ranks, directs the right flank officer to move, and to halt when proper; the left flank officer, at the same time, having these two objects to guide him, also places himself in that line—There are now three given points in a line, parallel to which, and at three paces distant the battalion should be dressed; the colour already ascertains the center of the battalion, and the sergeants covering the flank officers by taking three exact paces from those officers towards the battalion, determine its flanks and remain fixed.

When ordered to dress after halting.

fixed. The advanced officers now fall back, and having three fixed points in a line, confider the right as that of appui; *eyes* are ordered to be turned to the right, and they proceed inftantly to drefs their platoons as already directed.

It is evident, that in the dreffing of a fingle battalion after the halt, whatever correction is neceffary muft be made by advancing or retiring the flanks, and not by moving the center, which having been the guide in the march, has juftly ftopped at the point where it fhould arrive.

If the fingle battalion is not accuftomed at exercife when it is marching in front and halts, to be dreffed from its colours upon a given point beyond its flank (which may be a fergeant reprefenting the colour or flank of an adjoining battalion) it will never march or halt well, or be dreffed in a line when manœuvring with others; becaufe in fuch fituation, in order to have the juft line, each battalion muft be dreffed from its own center on the general line of colours.

---

When to incline.

Fig. 58.

When the battalion is to INCLINE or alter its length of ftep, the command muft be given in an audible and pointed manner, that the whole may at the fame inftant comply, and preferve uniformity of movement.

---

When to alter the direction of the march.

Fig. 60.

When in the courfe of marching, a fmall alteration of DIRECTION becomes neceffary—The advanced fergeants and colour, by bringing forward one fhoulder, and keeping back the other, infenfibly take it; the battalion, by fignal from the commanding officer, ftepping out on one flank, and fhortning the pace on the other, by degrees conform, new points of march are taken, and the whole again move on in front.

---

When the battalion is to RETIRE, it fhould previoufly be correctly dreffed, otherwife, its movement muft be difordered and unexact.

unexact. It is then ordered to *face* to the right about—At the command of *forward*, the sergeants and second colour, advance six paces from the rear rank, and the first colour moves into that rank—The lieutenant colonel leads, and the battalion moves on at the word *march*, in the same manner, as to the proper front; the music, drums, supernumerary officers, &c. taking care to march with exactness, not to interrupt, but rather to assist the battalion.

When to retire.

In marching by the rear, the battalion must cover its proper extent of ground—The rear rank men must not close their files too much, otherwise the front men, who are in general larger, will be crowded in their rank.

---

When the battalion is to FRONT, it receives the word: *battalion—halt*—to the right about *face*; and is then *dressed* in the manner already prescribed.

When to front.

---

When the battalion is marching in front, and partial OBSTACLES occur—Such divisions or platoons as are interrupted, *halt*, allow the rest of the battalion to move on, *face* outwards from their center, follow the adjoining flanks in *file*, and remain in this situation till they pass the obstacle, and can *move* up successively to their place in line which has been preserved for them.

When to pass an obstacle in advancing.<br>Fig. 61.

Should the wing of a battalion be obliged to break, it may then *march* off from one flank, by column of sixes or subdivisions, in order to prevent such an extension in file, as would otherwise take place; and it will again double up into line when the ground allows.

In this operation it is to be observed, that not even a single file should break without evident necessity—That no file should remain needlessly out of the line, but that as the ground allows they should successively advance into it—That either platoon or division breaks from its flanks, and two contiguous platoons of different divisions, must here

Attentions.

here confider themfelves as of the fame—That when a platoon is to break, behind which part of another is already broken, fuch platoon muft break, not to both, but intirely to one flank—That when the battalion halts, fuch part of it as is thus broken will front, and place itfelf fo as to flank the opening left.

Should the battalion be ordered to retire, while part of it remains thus broken, the broken part will *face* about, precede the battalion, not interrupt its movements, and re-enter the line gradually as the ground allows.

---

When to pafs through a wood.

When in the courfe of the march, the battalion is interrupted by a Wood, or fuch other difficult ground, as obliges the whole of it to break—The commander will give the word: *pafs by files to the front*; on which, each officer orders his platoon to the right *face*, *wheels* his leading file to the left, the reft follow, and he paffes as faft as the difficulties of the wood will allow him, endeavouring to keep a relative diftance from the right or left as ordered—When each officer comes to the edge of the wood, he orders his platoon to form up, and there remains with his back to the wood till the other platoons come out, and till he is ordered to *march* and drefs in the battalion.

---

To pafs an obftacle in retiring.

When the intire battalion is marching to the rear, if neceffary that part of it fhould break on account of any obftacle that interrupts the front, it will be done in the manner before directed—The line will continue in the ordinary march, but the parts which are obliged to break, will file off in time, and advance very quickly till they are before the part of the battalion which remains formed—They will then take care in moving on not to impede it, but will refume the ordinary ftep, and will by degrees re-enter the line when the ground allows of it.

---

Colours.

The colours muft always be carried uniformly and upright, in order to facilitate the moving and dreffing of the line.

The diſtances of battalions in line, are always taken from colours to colours; and the mounted officers in the rear, can beſt judge and apprize when ſuch diſtances are not correct. *Diſtances.*

When the WING platoons of a battalion are wheeled backward and faced outwards, in order to cover its flanks—Such platoons will march in file before the battalion during its retreat, and will take particular care to move in the ſame direction, and not to impede its progreſs—When the battalion fronts, the platoons will face outwards; and always recollecting that their immediate buſineſs is to cover the flanks, they will regulate their poſition and movements by thoſe of the battalion. *When the wing platoons form flanks to the battalion.*

*Fig. 59.*

When an intire battalion is to WHEEL (as to the right)—At the word: battalion to the right *wheel*, all the platoon officers place themſelves before the firſt file on the right of their platoons, and each platoon wheels up to the right one ſixteenth of the circle. At the word: *march*, the right platoon wheels into, and is halted in the intended direction: the others march ſeparately, join to the right, ſucceſſively enter the new line, and dreſs to the part firſt formed. Should at any time, a wheel in uniform front be required, it muſt be done gradually, the platoons waiting for each other, diſtances are preſerved from the ſtanding flank, and the degree of quickneſs is taken from the wheeling one—Should the battalion throw back a flank, ſuch part will face to the right about, take up its direction in the ſame manner by platoons, and then come about to the proper front. *When the battalion wheels on a flank forward or backward.*

*Fig. 63.*

There is another method, by which the flanks of the diviſions ſeem well protected during this operation, if made before an enemy—At the word: *march*, the right platoon wheels into

the

**Method of wheeling up or throwing back a division or the battalion.**

**Fig. 64.**

the new direction, and the rest of the line at the same time moves on in front, and *inclines* to join the left of the first platoon. When the right flank of the second platoon has arrived there, it also wheels up into the new direction, and the rest of the line continues the incline to join its left flank—in this manner the line preserves its uniform front, inclines, and gradually enters the new position as its leading platoon arrives in it, at the same time that it covers and protects the flanks of the formed platoons. By the same means also will a battalion or line throw back any number of its divisions in presence of an enemy. The angular platoon will give the direction, the rest will *face* about, *march*, *incline*, successively *wheel* into it, and *front*. The outer platoon of all which is supposed formed as a flank to the battalion, will march in *file*, and cover the flank.

---

**Passage of lines.**

**Fig. 65. A.**

In narrow grounds, where there are redoubled lines, and and in many other situations, it becomes necessary for one battalion to pass directly through another in marching either to front or rear; but this must particularly happen when a first line is to retire through, and make place for a second.

**Passage of first line through the second.**

Should the second line be ordered to advance, and occupy the ground which the first is to quit——As soon as it approaches within twenty paces, and halts, the front line battalion receives the word: *pass*—Each platoon is ordered to *face* to the right, and disengage its head—Each platoon *marches* at a quick step to the rear; passes straight through the second line, and when past resumes the ordinary march.

**Fig. 65. A.**

Wherever the heads of the retreating files present themselves, the officers of the second line cause four files of their platoons to fall back, and again to resume their places when the others have passed.

**Reforming of first line.**

**Fig. 65. B.**

The officers of the first line having been cautioned again to form at (one hundred and fifty or two hundred) paces from the second, and having begun to take and count them from the passing of that line, will *halt* accordingly; *close* their platoons; *front*; place themselves on the pivot (left) flanks; take their just distances, and aligne

aligne to the front on the pivots of the three first platoons, which will be instantly arranged by the major. When the whole is in order, the battalion is formed by the *wheel* of platoons to the left.

Counting the paces, which must be of the accurate regulated length, is an assistance for the halt, rather than the exact means of giving the alignement, which must depend on the quick, and after dressing of the pivots from the front.

If the second line remains posted—The first retires in front, till within twenty paces of the second—at the word *pass*, each officer *turns* his platoon to the left; *marches* quick in file through the second; resumes the ordinary march when past, and *halts* at his determined number of paces from that line.

Passage of first line.

Fig. 65. B.

---

Should a new position not parallel, be taken by the passing battalion—The commander with his two leading platoons will first enter it, and direct the others to regulate their flanks by them; and if several battalions are passing the second line, the alignement is thus made easier for them.

First line reforms in an oblique position.

Fig. 65. C.

---

Where a HEIGHT in the rear is to be crowned by a retiring line—an officer must not dress exactly to the platoon that precedes him; but in joining it, he must halt, and arrange his own in such a manner, that the slope of the rising can be entirely seen and commanded, which is here the great object, and not to be attained if the troops were to adhere to a straight line.

Crowning a height.

Fig. 65. E.

---

It may happen where the passing line is to post one flank, and refuse the other, that the officers take their distances from behind; but if this movement is foreseen, the inconvenience may be avoided by filing from the proper flank: for if the left is to be posted, and the right refused, the platoons may *pass* from their left, the column will thereby have its left in front, will be more readily directed on the point d'appui, and the preservation of distances will be facilitated, as they

Refusing a flank.

Fig. 65. D.

they will then be taken from the front. If the right is to be posted, the platoons may pass from their right.

---

## MOVEMENTS of the LINE in FRONT.

<small>Chief object of movement.</small>

The chief object of every other movement, is the quick and just formation into Line when necessary, and the consequent march of that line in front towards the enemy, which is the last great operation that precedes the attack, and is the most difficult and material of all manœuvres.

<small>Necessity of order.</small>

No body of troops can advance in line on an enemy but in disorder and confusion, unless the original formation of that line has been perfectly straight; and its correct preservation during the the march requires every attention—While order subsists, the soldier feels his advantage, exerts himself, and acts with energy and spirit—When disorder prevails he perceives his inferiority, desponds, loses all confidence in himself or commander, personal safety soon occurs, and the moment of flight is not far distant—To bring up troops to the attack in imperfect order, is to lose every advantage which discipline proposes, and to present them to the enemy in that very state to which after his best efforts he hopes to reduce them.

When a line of several battlions is formed and halted, there is an interval of twelve paces for two pieces of artillery betwixt each, and the whole generally are dressed to a flank. The formation and march of the battalion in front have been already detailed, the same principles direct that of the line.

When the line is to march in front, one of the battalions is *named* as the regulating one to whose movements all the rest are to conform

form—At the caution given by the commander of each battalion that the line will move *forward*, the front colour and serjeants of each will carefully move out three paces as already directed, and the perpendicular point of march will be given to the regulating battalion only, by the major in the rear—The advanced colours will take the utmost care in moving out to dress by the regulating one, so that the whole may be in a line perfectly parallel to that on which the corps is formed, and which cannot be otherwise, if they are accurate in taking their three paces. And, should the line at any time be ill dressed when halted; the advanced colours must still be correctly placed before the line is again put in motion, and the several battalions in advancing will conform to them.

March of the line in front.

At the word *march*, rapidly repeated, each battalion at the same instant is put in motion, dressing to their several centers.

Every attention required in the movement of the single battalion must be redoubled in that of the line: on the regulating one particularly does every thing depend, whose pace must be steady, uniform and direct—The commander of the line who is with it, must from its advanced colour observe and so caution an adjoining one, that the others being enabled at all instants to dress with these, may preserve and carry on a line perfectly parallel to the original line; and the several battalions by scrupulously following and conforming to their leading colours, will if these move with justness, advance on the most uniform front.

Attentions in marching.

The difficulty of ascertaining the perpendiculars of march to the regulating battalion are certainly great, as the least deviation to right or left will tend to alter the direction of, or to disorder the line; but there appears no other marked general rule that can be given, and the greatest care therefore must be taken in remarking them; it is from the *rear* of the body only that they can justly be observed and announced, and it is from thence that any improper deviation in the march can best be corrected; much therefore depends on the attention of the majors and adjutants.—It would seem that the allowing each battalion to take its distinct points should tend to preserve the uniformity of the line; but it is much better to make the others depend on the movement of the regulating one, and on the parallel march of the advanced colours, than in any shape to trust to their own points which will never be true, with respect to each other, and from whence a crowding in, or opening out must infallibly arise.

Should any battalion be behind the line—at the word *advance*, the step will be lengthned till it arrives in line.

Should any battalion be advanced before the line—at the word *mark*, the step will be shortned till the line comes up.

When necessary (in order to correct intervals) battalions will occasionally *incline* but without heads being turned from the center, and when the intention is answered they will receive the word *front* to cease inclining.

<small>Attentions in marching.</small>

In correcting the movements of battalions in the line much judgement must be exercised, the general situation must be examined, inclining must not take place on every small alteration of interval; and wherever the fault does originate, the remedy should in general from thence begin. The opening or closing of intervals commonly proceeds from the colours moving in a false direction, and should be corrected by their taking a true one.

The march in line of a considerable body, can only be at the ordinary step, a quicker movement would produce disorder, nor could artillery attend its movements when advancing to the enemy. But there are situations where a brigade or smaller front should move on to a particular attack at a lengthy step, or where even a quicker cadence may be required from them.

The commanders of battalions must look upon the regulating one as infallible; and the general must watch over it, and carefully direct its motions. The *march*, and *halt*, and attention of each battalion in line, is by its own center; the commander alone attends to the regulating one. *Dressing* to a flank is by a separate direction, given when necessary or proper after halting.

The majors and adjutants, will often place themselves close to their third rank, and to their second colour, to see that they are aligned with their own battalion and the next, and that they are not too much advanced or retired—That eyes remain fixed on the advanced sergeants and colour—That those do not look behind them or turn their heads, and that the center colour observes its distance of three paces.

*Attentions in marching.*

---

Battalions in line, marching over heights or across vallies, will require more time to pass them, than others who are moving on the same extent, but of level ground; in order to preserve equality of front, the last must therefore shorten their step. The same thing will happen, when platoons, and particularly the colours, are obliged to break and fall into file. The majors and adjutants who attend the march, will easily see those who are advanced or retired, and correct them accordingly.

*Attention on uneven ground.*

The flanks of battalions, must take particular care never to advance before the center; the officers of the wings, by attending to the general line of the colours, will much assist; and if those officers and the center, can be preserved in the true direction, the internal dressing of the battalion will readily follow.

*Flanks of battalions.*

In line—Cautions and words of command, are repeated by commandants of battalions only; but in such manner, as to be heard distinctly by the leader of the adjoining one, beginning from that which regulates. The words *march—halt*, &c. of each, must instantaneously be circulated.

*Directing battalion.*

The battalion which is nearest to, and is to preserve the point of appui, will in general be the regulating one; therefore, a flank battalion will commonly direct the movements of the line, and should the commander change it, he must announce such change to the line.

---

The line thus marching in perfect order, when it arrives at its destined point, receives the word *battalion*, rapidly repeated

When<br>marching<br>and to halt.

to each flank; the colours of the line endeavour to dress by each other and the regulating one, and cease to advance; their battalions which are behind them, and at their three paces distance, having such an evident direction, *mark* the step, and *dress* within themselves. After a pause of five or six seconds, the word *halt* is more instantly circulated if possible, and each individual battalion halting in the manner already prescribed, the general alignement should be nearly just.

The whole line must *halt* at the same instant when the word is given, and no dressing or correction of intervals takes place till so directed—The advanced colours on the halt fall back to the battalions.

---

When halted and ordered to dress.<br><br>Fig. 66.

But, if the commander finds it necessary to give a more exact dressing, he immediately orders the first colour of the regulating battalion, again to move forward three paces (its place being supplied by the second one) and also a colour from one of the two adjoining battalions—Being himself then placed at the regulating colour, he directs the position of the other advanced one; their flank officers also dress by them; thus, the colours and flank officers of those two battalions are in an exact direction, parallel to which, and at three paces distant the line of troops is to be formed. The other colour immediately takes its distance from its respective advanced one, the covering sergeants mark the flank points of their battalion, and the correction of dressing instantly begins in the manner already prescribed. The other battalions of the line do not advance their front colours, but dress them exactly on the second colours of the two directing battalions; and their flank officers having now so many fixed and visible points ascertained, can easily place themselves in the true line, and begin the correction of their battalions—When the line is dressed, the two advanced colours resume their proper station.

Attentions in dressing.

If when the advanced colours move forward, the rest of those of the line are immediately lowered, and not again held up to view, till the persons of their bearers are successively

cessively placed on the true line, it will be a great assistance to the distant battalions.

The difficulty of distinguishing and directing in an ill formed line, would make it troublesome at once to give a true position to the colours and flank officers, without first procuring the advanced parallel line as above prescribed.

The dressing of the line may appear tedious in the detail; but the alertness and intelligence of individuals will make it sufficiently quick in practice considering the difficulty of the operation and the great object to be attained, if each attends solely to his own business, and does not improperly intermeddle in the province of another—For if the commander of the line will limit himself to giving the alignement to the colours of two battalions, and only watch over the others——If those of battalions attend chiefly to the position of their colours; if the flank officers take their points as soon as their colours are arranged; and if the platoon officers can be depended upon in their several situations, every thing will go on quick; and in proportion as the colours of each battalion are aligned, will their several component members be truly placed. On the same principles that the dressing is corrected, is the march in line to the front conducted.

If the line is at any time ordered to *halt*, without being correctly dressed, and is immediately after directed to resume its march—As soon as the colours move out, they will be instantly placed in the same line from the one regulating, although by this means, some of them may be more than three paces distant from their proper battalions. At the word *march*, they will move on preserving this line, and the battalions will gradually attain their three paces from them. This will take place whether the line is to advance or retire.

---

When the line is to RETIRE, the necessity of its being previously correctly dressed, is still more essential than it

*When the line is to retire.*

it was in the single battalion; if that preliminary is not granted, its movements must be disordered in proportion to its extent. But as there may not always be time to give it the wished for degree of exactness, before the retreat begins; such aids may be applied, as will greatly assist it in the course of its movement.

The battalions are ordered, to the right about *face*—The sergeants and second colour of each, at the word *forward*, advance six paces from the rear rank, the first colour moves into that rank, and the point of march is given from the rear to the regulating battalion—The advanced colours, which are probably not in an exact line, are immediately so placed from the regulating one, without regard being had to their distance from their respective battalions, and an exact line of colours is thereby obtained by the majors and adjutants. The whole is now ordered to *march*; the advanced colours preserve their position and line, and the several battalions by degrees, acquire their just distance of six paces from them—The lieutenant colonels lead, and the line moves on in the same manner as to the proper front. The music, drums, supernumerary officers, &c. take care to march with exactness, not to interrupt, but rather to assist their battalions.

<small>When to front.</small>

When the line is to front, each battalion receives the words, *battalion*—*halt*—right about *face*; and if it is then to move forward, the colours and sergeants immediately advance before the front rank, and are there correctly lined, ready to conduct its march. But it is to remain halted; the dressing is then ordered in the manner already prescribed.

<small>Attentions in firing on the march.</small>

In the march of the line, either to front or rear, when parts of it (as battalions or their wings) are ordered to halt and fire—The general step must be much shortned, to allow those that fire to regain the line without confusion, and the attention of the advanced colours to regulate the whole must be very great.

What-

Whatever impediments present themselves to the march of the line in front, will be avoided by the methods prescribed for the battalion; and the openings made by such parts as are obliged to quit the line, will be carefully preserved, in order to their re-entering into it, as soon as the ground permits.

*Passing obstacles.*

---

INCLINING the line when well performed, is a most important and decisive operation—It enables in advancing to gain the flank of an enemy, or to preserve the appui of a river, wood, &c. the line of which runs oblique to that of the march—The principles of this diagonal movement have been already explained, and the difficulty of carrying on the line parallel to its first position; therefore, the perfect squareness of the shoulder and body of each individual to the front, is the first great object; and every attention must be given to the advanced colours, that they preserve their parallel line, and remain a true direction for their several battalions. The whole line look to the flank to which they incline, the step must be correct and uniform, and the attention of every individual on the stretch. The Prussian troops alone can perform this movement in line with success and correctness, and it is of all others the most difficult.

*Inclining.*

*Fig. 66.*

---

In the autumn of 1785, there were many opportunities of seeing the movements of the PRUSSIAN troops in line. Among others, at MAGDEBURG, eighteen battalions extending about two thousand yards, entered the alignement in open column; formed in line; marched in front in the most perfect manner above one thousand five hundred paces; and during the course of the march, fired by battalions advancing. The line afterwards retired a considerable part of that distance.——At BERLIN, the whole infantry of the garrison, consisting of eighteen battalions, marched in open column in the alignement; formed exactly in line by the signal of a cannon; marched in front about one thousand paces; and

*Prussian movements in line.*

*Fig. 67.*

and fired while advancing by divisions and battalions.——In SILESIA the infantry was generally formed in two lines, of eleven thousand men each, and the lines frequently marched in front separately and well for fifteen hundred paces. The two lines, on one occasion, notwithstanding an exceeding heavy rain the greatest part of the time, and directly in their faces, marched, and preserved a most exact line, both in advancing and retiring, for above half an hour; and in retiring the whole infantry obliqued to the left, and in the course of the march gained at least the front of two battalions to that flank.——At MAGDEBURG, two lines of ten battalions, which had marched off in columns of wings, and were halted in close columns on the *alignement* of each line, ready to *deploy*, moved forward about three hundred paces, and at the same time obliqued to the right: they then halted, and *deployed* into line.

Fig. 68.

---

# ECHELLON.

General application of the echellon.

By an oblique march the enemy may have been approached, even outwinged, and the following flank may have been refused to him, and kept back from his fire; but it is evident, that he cannot be reached without exposing the leading flank, unless the whole line inclines, which is a movement subject to great difficulty on so serious an occasion. The attack in ECHELLON is therefore substituted—which preserves the advantages gained by the oblique march, has not the faults to which the incline is liable, aims at the weak part of the enemy, does not risk the whole of your force, is particularly applicable to the movements of infantry, or the mixed ones of cavalry and infantry, more to those of great bodies than small, and equally whether the object is to advance, or retire.

In

In advancing, the several bodies move independent, act freely, and are ready to assist. In retiring, they fall back upon each other, and thereby give mutual aid and support; nor can an enemy press the pursuit but with much circumspection, and without endangering his own flank.

---

In *echellon*, the divisions are retired at equal but parallel distances behind each other, and in that situation they are ready—to move up into parallel line—to wheel into oblique line—to file into column— to retreat upon the rear, or any other division. Fig. 73.

---

If the object is to form in parallel line to the front—whatever distance the divisions are retired, the flank of each following one, must cover that of the one preceding it. Fig. 74.

---

But when the object is to form in oblique line—in proportion as the divisions are retired, they must cover part of the one preceding them, viz.

If the divisions are retired one fourth of their front—they must cover a twenty-eighth part of the preceding one; and the oblique line will then make an angle of about fifteen degrees with the parallel line. Fig. 75.

If the divisions are retired half of their front—they must cover one seventh of the preceding one; and the oblique line will make an angle of about thirty degrees with the parallel line.

If the divisions are retired three fourths of their front—they must cover one third of the preceding one; the oblique line will then make an angle of forty-nine degrees with the parallel line.

In this manner a proportionate allowance must be made for any other intermediate distance, remarking that the nearer the oblique line of the *echellon* is to a right angle with the parallel line, the greater must be the distance betwixt divisions, and the greater proportion of front must each cover.

If the divisions are retired the whole distance of their front, they will then completely cover each other, and become an open column, ready to form in line at right angles to the front of the march.

---

§. 73.

Whatever diftance the divifions are retired behind each other, if thofe diftances are equal, and the divifions alfo equal, the flank files will always line in a diagonal direction—This muft be attended to, as an effential guide and rule; and therefore when the fecond divifion has at any time taken its pofition, the following ones will the more eafily take up theirs, by obferving the line of the flanks, and their prolongation.

---

The diftance that divifions are to take, muft always be announced, as well as the intention; whether to be prepared for forming in parallel or oblique line, that they may regulate their flanks accordingly.

---

When the *echellon* is unconnected with a line, the advanced flank regulates all its movements; when attached to a line, it muft depend on the motions of that line.

---

The whole, or only part of the body may be thrown into echellon, and that either to the front or rear. In the firft cafe, with a view to gain the flank of an enemy, or obtain a crofs fire: in the fecond, to refufe, or to cover one's own flank.

---

The firft pofition of the deployment of the column is always a pofition in echellon.

The battalion may form echellon, either in whole or in part, on any central divifion—Either marching or halted—to front or rear, and from either flank.

---

Where a full line has to advance in front—To prevent the crowding and breaking, which want of intervals might help to occasion, it may readily take, and advance in a degree of echellon position. Each front rank retired at least as far as the preceding one's rear rank. The battalion of direction is the leading one, which must move with the greatest exactness, and when so ordered, the whole can in an instant move up into line. Fig. 77.

---

Where a line is passing a defilè to its front, and from its center— It may first form at the head of it in the echellon *position*: the several divisions are then ready to move up into line; or by wheels towards the flanks, to form in oblique lines and protect those flanks. Fig. 78.

It may also in the same manner pass a defile to the rear, retiring from the flanks by echellon, while the center protects the movement.

| Nature of Movement | Commands given by leaders of | | Circumstances of Execution |
| --- | --- | --- | --- |
| | Battalions | Divisions | |
| When the battalion is marching, and the right wing is to form echellon to the rear. Fig. 79. | *The right wing of the battalion will form echellon to the rear on the seventh division at half distance—to form oblique line.* | | ——— A caution. |
| | | *Halt* | The six right divisions, the rest of the line moving on. |
| | | *Left, march* *Left, incline* *Forward* | The sixth division, when the line is at the prescribed distance from it, will—march—incline, till it has covered one seventh of its preceding division, and then follow: and thus successively, every other division to its right. |
| | *Halt* | *Halt* | The line and echellon will halt when ordered. |

|  | Commands given by leaders of | | |
|---|---|---|---|
| Nature of Movement | Battalions | Divisions | Circumstances of Execution |
| When halted, if the echellon is ordered to form oblique line to the right.<br><br>Fig. 80. | The divisions of the echellon will wheel to the right into oblique line. | | ——— A caution. |
| | | Right Wheel | The pivot man of each division faces to the right into the intended oblique line. |
| | | March | Each division wheels to the right into oblique line. |
| | | Halt | Each division when in the oblique line, and dress to the right. |
| When the oblique wing is to resume echellon.<br><br>Fig. 81. | The divisions will again form in echellon. | | ——— A caution. |
| | | Left, wheel | The pivot man of each division faces square to the front of the line. |
| | | March | Each division wheels up to the left. |
| | | Halt | Each division when parallel to the front of the line. |

| Nature of Movements | Commands given by leaders of | | Circumstances of Execution |
| --- | --- | --- | --- |
| | Battalions | Divisions | |
| When the echellon, halted and formed to the rear, is ordered to advance and form echellon to the front on the same division. Fig. 82. | *The echellon will advance, and be formed to the front.* | | — A caution. |
| | | *March* | The whole divisions of the echellon move forward; when the division next the standing part of the line arrives at it, such part of that division as is obstructed by the line will fall into file, in that shape pass it, and then resume its place when in front of it. |
| | | *Halt* | The division next the standing part of the line will halt when exactly at the distance in front, and in the same relative situation as when in rear of it. |
| | | | Thus successively each division will pass the one preceding it, and the right division of the whole will be now the advanced part of the echellon. |
| | *Front to the flank.* | | In this situation by a wheel to the left, the echellon will flank the front of the line. |

| | | |
|---|---|---|
| When the echellon is to return into the general line which is halted. | If the echellon formed to the rear, is to move up into line. | The several divisions *march, incline, halt*, and *dress* successively to the standing part of the line. |
| | If the echellon formed to the front, is to fall back into line. | The several divisions *face* about—*march* to the rear—*incline*—*halt*, in line—*face* about, and *dress* successively to the standing part of the line. |

| | | |
|---|---|---|
| If the line is in march to the front, when the echellon is ordered to join it. | If the echellon formed to the rear, is to move up into line. | The several divisions will increase their pace; *incline* outward if necessary, and *move* up successively into line. |
| | If the echellon formed to the front, is to join the line. | The echellon will *halt*. As the line continues to advance, each division successively will *march* in due time, *incline* outward as far as is necessary to acquire sufficient ground—*front*, join the line when it comes up, and proceed with it. |
| | | The same operation but reversed would take place, should the line be retiring, when the echellon is ordered to join it. |

When the battalion is halted, and the right wing is to form in echellon, either to front or rear.

- If to the front: The intended distance of divisions is named—The divisions of the right wing are then ordered to *march* forward—When the division next the standing part of the battalion has acquired the prescribed distance it *halts*. The others continue moving on, and halt successively in echellon, the whole dressing to the left.

- If to the rear: The intended distance of divisions is named. The right wing goes to the right about—*marches*—*halts* successively each division—*fronts*, and *dresses* to the left, the whole being then in echellon to the rear.

After the echellon is halted, if the object is to be in readiness to form the oblique line.

- The whole divisions must separately *face*—*march* to the left—*cover* each other in proportion to their prescribed distance, and again *front*.

- The same operations take place whether the whole, or part of the battalion, or line, form in echellon.

The ECHELLON position, is a very advantageous manner of making an attack, with the center or one of the wings reinforced—If successful, the divisions move up into line to improve the advantage—If repulsed, they are in a good situation to favour and protect the retreat.

Different previous manœuvres must always have distracted and divided the attention of the enemy, and prevented him from being certain of where the attack is to be made in the echellon position.

With respect to the enemy and the intended march, it may be taken from the - $\begin{Bmatrix} \text{parallel} \\ \text{oblique} \\ \text{column.} \end{Bmatrix}$ position.

If from the parallel line—It is previously divided into the several *echellons* which compose it; and the distance at which they are to remain behind each other, is announced.

**Attack in echellon from the parallel line.**

The reinforced flank, which is to attack, is then ordered to advance—Each echellon of two or more battalions moves on, when the preceding one has gained the ordered distance (of perhaps one hundred yards) and thus being regulated by the head, act according to the event of the attack.

Fig. 89.

Fig. 90.

Though the inclining with a considerable line, is hardly to be attempted when near the enemy—yet, that operation may be performed by the *echellons* in advancing, so as to gain ground to the flank; but it requires every circumspection and attention to preserve the relative positions, and in time to steady and resume the direct front. All such movements must be critical near the enemy, and in general his flank will be best gained by troops placed in column for that purpose, behind the attacking body, and which are there ready to form up and improve such advantages as offer.

If from the oblique line—This position having been taken from the column of march, or in the course of advancing in line upon the enemy; the attack will be made either by the advanced or retired flank reinforced; the divisions of the line, and the distances of the *echellons* being previously ascertained.

When

When by the right flank advanced—Each two or more battalions wheel up to the right, as much as brings them parallel to the enemy's front; the whole may then be confidered as fo many diftinct, but perfectly parallel lines—The attacking echellon moves on, the others alfo march fucceffively, when their preceding ones have acquired the ordered diftance.

When by the right flank retired—The right *echellon* will be wheeled to the left, till placed parallel to the enemy's front; the echellon next to it, will alfo wheel up in the fame manner—The attacking echellon will then march, pafs the other, and proceed: the fecond echellon will alfo move on, when the given diftance is acquired; and the third echellon will at that time wheel up to its parallel pofition, and advance in its turn. In this manner, they will fucceed each other till the whole are in the fame fituation, as when marched off from an advanced flank.

The advanced echellon being arrived at its object, the attack begins, and the others attend the event—If it fucceeds, they move up into line to perfect it—If it fails, each falling back on each is ftrengthned and fupported every inftant of the retreat.

The enemy's cavalry will hardly at any time attempt to enter the intervals, as they muft always give a flank to an advancing echellon.

*Fig. 83.*

*Attack in echellon from the oblique line.*

*Fig. 84.*

---

In order to place the leading *echellon* which becomes the guide for the others, perfectly perpendicular to the line on which it is to march up to the attack—The commander muft himfelf take the greateft care to wheel up its flank divifion, exactly into that direction, and whofe prolongation is the line on which the whole echellon is to be formed. When this divifion is accurately dreffed, the number of paces that its wheeling flank has taken, are meafured and intimated (as well as the ftrength of the divifion if neceffary) to each of the other echellons; and their equal flank divifions wheel up in the fame manner, fo that the outward wheeler fhall halt at the given number of paces, which are previoufly marked for him—Thefe parallel divifions, become each a direction

*Attentions in forming the echellon.*

tion for its own echellon, their several fronts are prolonged, and all the other divisions and battalions march up, and dress with them.

The larger the original division that thus gives the direction, is, the more exact will be the first formation of the several echellons; a half battalion may therefore be thus placed. The divisions of direction must be of equal strength, otherwise the number of paces will not be a true guide. The paces must be of correct length, and taken in the chord of the wheel.

---

Notwithstanding these measures to obtain exact parallel lines, the following *echellons* must on the march be guided by, and conform to the leading one; their great object is to preserve in moving on, their parallel and relative situation. In this they are to act in the same manner, as when advancing in line; and having the leading echellon to guide them, together with the assistance of the mounted officers who attend to their movements, and prevent their outward flanks from being thrown too forward, they will execute with justness this important manœuvre. The preservation of intervals is also as essential an attention, as in the attack in line.

Attentions in forming the echellon. When the leading echellon is halted and the others are to move on to line with it; being then within reach of the enemy's fire, they must take particular care not to throw forward their outward flank, but rather to refuse it—This can be accomplished by the advanced colours taking heed to halt exactly in the true line of the attacking echellon; and it depends on the conductor alone of that attack, to give it such a direction, that its prolongation shall pass before the enemy's front.

---

The second line when there is one, follows in every thing the movement of the first—The battalions make the same wheel, preserve the same relative position, and serve as a support to the

first

first—The attack of the second line, therefore, moves on at the same time with that which it is to support. The echellons of one or more lines, are generally retired from one hundred to one hundred and fifty paces, each behind the one preceding it of its own line.

---

The *echellon* may be formed from column. The heads of the columns are halted at given points, and given distances—The attacking bodies form in one or more lines, the others extend to the flank in echellon, being separated, a space equal to the distances they halted at in column—This space is augmented if necessary, when the whole move on, and lines of two or more battalions each are formed. From the echellon position by flank marching, the order of column may again be resumed.

Echellon formed from column.

Fig. 85.

---

A central attack may be conducted in this manner—Three columns are directed towards the central point, the middle one forms in lines and makes the attack, the others in echellon position sustain it.

Central attack in echellon formed from column.

Fig. 86.

---

The enemy are posted in A, the body is formed in line B, with a reserve of three squadrons; and two squadrons are placed in column, destined to turn the flank—The whole by inclining gain the position C; from which the echellon position D is taken, and the attack made on the enemy's left.

Movements in line and echellon.

Fig. 89.

---

In SILESIA in 1785—The line of infantry of eleven thousand men, supported by fifteen squadrons of cavalry (each covering a battalion) in second line, and ten squadrons on each flank, when about six hundred yards from an enemy advanced to the attack from the left in four echellons, retired about one hundred and fifty paces behind each other—The attack being repulsed, the echellons

In Silesia.

Fig. 90.

fell back upon one another, and upon the right flank which had remained posted on a favourable height.

<small>At Potsdam.

Fig. 87.

Prussian movements in echellon.</small>

At POTSDAM, two lines of infantry of nine battalions each, moved from the right on to the attack, each line in three echellons, each echellon of three battalions, and retired about two hundred paces behind one another; a third line of cavalry also supported—The attack was repulsed, the left remained posted, and the echellons of the two lines retired alternately through each other, the right of the whole being gradually much thrown back during this operation.

<small>At Magdeburg.</small>

At MAGDEBURG, two lines of nine battalions each, advanced from the left in three echellons of three battalions each, and retired about eighty paces behind one other. In this manner, they marched with the most perfect correctness about one thousand one hundred paces, when a new disposition was ordered to take place.

---

The cannon of the posted echellon, can always procure an enfilading fire, on whatever body is pressing on the retiring ones, and will thereby very much assist the retreat, and check the pursuit.

---

## Of Second Lines.

THE movements of a SECOND LINE, are similar to those of the first—It will always preserve its parallelism and distance.

<small>Fig. 91.</small>

Its front is regulated by its own division or battalion of direction, which moves by that of the first line.

In changes of position, its movements correspond to those of the first line.

In forming the line, it will march upon its own points which are parallel to, and ascertained in consequence of those of the first. Fig. 92.

If the first line makes a central change of position—the second line must make a change upon its corresponding division to comply with it. Fig. 93.

The movements of *reserves* are similar to those of a second line, being determined and directed in consequence of those of the first line. Fig. 91.

When the lines break in column to the front—the second will generally follow the first.
When the columns are formed to the flanks—the second line will often compose a separate column or columns. Fig. 94.

When the march is to the rear—the second line will often lead in column.

As the movements and situations of second lines, are dependent on those of the first—To prevent interfering, they must be regulated by certain general rules, which are applicable according to circumstances.

General rules in the movements of second lines.

Although the lines may execute all their movements, by forming columns of battalion, regiment, brigade, or even of greater numbers—yet, whenever the columns of the preceding line shall exceed in depth, the distance between the two lines, the following line will be often retarded in its movements forward, till the surplus

surplus of such column has marched over a space equal to its length.

Hence is apparent the necessity of multiplying columns, which become manageable, and acquire precision in proportion, as their depth is diminished.

The distance betwixt the lines, may be supposed equal at least to the front of two battalions and an interval.

---

Fig. 95.

Whenever the new line of direction passes through the extremity of the right or left, and that the front is to be outwards; the movements of the first line would be interrupted, if the second was to march by its flank—The second must therefore make a degree of central movement, and the center point in that line, will be distant from the flank, in proportion to the fixed distance between the lines, and to the angle more or less open which the new position makes with the old one.

---

Fig. 98.

The same principles determine in central movements of the lines— For example, if to face to the right the more or less open, that the angle formed between the old and new position is, the more or less must the central point of movement for the second line be looked for towards the left—The reverse takes place where the lines are to face to the left.

General rules in the movements of second lines.

When the direction of the new position of the first line is known, that of the second line will be equally so, as the due distance is already determined.

The point which ought to be taken for a center in the second line in central movements will be also known, since it is always that, through which passes the prolongation of the new direction of the second line.

Fig. 96.

In a central movement to the rear, as a second line in that case, necessarily regulates the movement of the first—The first will observe with regard to the second, what in other cases the second observes with regard to the first.

When-

Whenever the first line breaks and manœuvres by its right to face to the left, or by its left to face the right—the movements of the second line are free and unembarrassed, and it may turn round the manœuvring wing of the first line, and take its new position behind it, by extending itself parallel to its direction how oblique soever that may be; and in the above cases, they enter into the new position before the line of direction—This movement is equivalent to a battalion changing its front, to one flank, by filing from the other. *Fig. 99.*

The central movement often required from the second line, to conform to that of the first, is equivalent to that line marching in two columns of divisions from near the center, and from that situation forming to either flank.

The movements of the central columns well understood—Those of the battalions of the wings are similar in the two lines. *General rules in the movements of second lines.*

In central movements to the right, and either made on a fixed point, or one moving to front or rear—The battalions of the right wings of both lines arrive with the left in front before the new line of direction; and the battalions of the left wings of both lines arrive with the right in front behind the new line of direction.

Second lines are seldom composed of as many battalions as the first—They are often divided into distinct bodies, covering separate parts of the first line.

Second lines will not always remain extended, they will often be formed in column of battalions or of greater numbers, ready to be moved to any point, where their assistance is necessary.

Where several lines of attack are formed, the second should always out-flank the first, the third out-flank the second, &c. the advanced one being thereby strengthned and protected. *Lines out-flank each other.*

The officer commanding the second line must always be properly informed of the nature of the change to be made by the first, that he may readily determine his corresponding movements.

It requires much attention—To conduct heads of battalion columns of both lines, nearly parallel to their lateral ones, and perpendicularly, or diagonally to front or rear, according to the nature

ture of the movement——To determine with precision and in due time, their points in the new line, that wavering and uncertainty of march may be avoided—In great movements to allow the soldier every facility of motion, without increasing the distances of divisions, and to require the most exact attention on entering the new line and in forming—To avoid obstacles in the course of marching, but as soon as possible to re-enter the proper path of the column— While out of that path, the leading colours of that column may be lowered (as a mark for the neighbouring column not to be then intirely regulated by it) and again advanced when it regains its proper situation.

*Attentions in the movements of second lines.*

Fig. 100. Different NEW POSITIONS of TWO LINES, taken from the ORIGINAL POSITION.

If to face to

No. 1.

A    BOTH lines will break to either flank, and advance in columns of battalion to the new position.

Both lines will break to either flank and advance in columns of battalions to the new position; the second line will there pass beyond the first; when formed the lines will be inverted, but not the battalions within themselves—

O    This inversion of the lines can be avoided by countermarching both lines before the movement—Or by countermarching the battalions and then the lines after the movement—Or by marching from the old to the new position from either flank.

*Positions of second lines.*

2. The

If to
face to

### 2.

A    The lines will break to the right, and march diagonally in column of battalions to the new pofition.

O    The lines will break to the right, the heads of them will be pointed towards the right flank of the new pofition, and when they arrive in it, and halt, the battalion columns will lead off to the right (the rear line preceding) and gain their places in the new pofition.

### 3.

B    The lines will break to the right, march on to the interfection of the new lines, and then along them to the right; the rear of the columns entering the new pofition by battalions when the front halts—Or—The lines will break to the right; the leading battalions will point to their places in the new line; the others will lead off to the left diagonally; the whole will enter and form in the new line.

Pofitions of fecond lines.
Fig. 100.

O    The lines will break to the right and either continue their march to the point of interfection—or—The leading battalions be directed to the right flank of the new line, and the rear battalions of both lines difengage to the right.—The fecond line will precede the firft, and with freedom manœuvre round its right flank.

### 4.

B    The lines will break to the right and march in column to any point towards the right of the new line, which they will there enter, and move along. The rear battalions when ordered will difengage to the left, and take their places in the new line.

O    The lines will break to the right—The battalion columns will lead off diagonally to the right (the fecond line preceding)

to

( 162 )

to their points in the new position, being regulated by the right battalions.

### 5.

O  The same operation as in No. 1 facing A, except that both lines retire, and that the second precedes.

C  The same operation as in No. 1 facing O, except that both lines retire, the second preceding—When the second line is in the new position, the first must pass beyond it. The same methods to remedy inversion will also take place.

### 6.

Positions of second lines.

D  The lines will break to the left, be directed to the left and rear into the new position, and march along it, till they arrive at the left.

Fig. 100.

C  The lines will break to the right, be directed upon the left of the new position, enter it, and march along it.

### 7.

D  The lines will break to the left, be directed upon the right of the new position, and march along it to the left.

C  It is a countermarch of the lines in column from the right (the first line round the second) and then a prolongation of the lines to their new position.

### 8.

D  
O  } The same as No. 4. but performed to the left.

### 9.

A  The lines will break to the left, and prolong their march to the new position.

D  The lines will break to the left—Prolong their march to the left of the new position—The rear battalions will then

If to
face *to*

then disengage diagonally to the right, and take their places in the new line—The second following the movements of the first line.

### 10.

A  Both lines will break to the left, the heads of the columns will be pointed towards the center of the position, and when they arrive in it, will march along it and form.

D  Both lines will break to the right, and by a kind of countermarch, enter into the new direction.

### 11.

D  Both lines will break and make a central movement to the rear, on the divisions where the new line intersects the old, the second line preceding—When those divisions halt, the columns of battalions will throw themselves into the new line.

O  Both lines will break and make a central movement to the front, on the intersecting divisions. When those divisions are at their ground, the columns of battalions will throw themselves into the new line.

Positions of second lines.

Fig. 100.

### 12.

B
O  The central movements, are the same to front and rear for both lines, as prescribed in No. 11—The right flanks leading.

### 13.

B
D  Both lines make a central movement to the rear, on the intersecting divisions—When at their ground, the columns of battalions take their places in the new line.

( 164 )

If to face to

#### 14.

B     The lines break to the right—the columns of battalions march diagonally to the left into the new position—the second line conforms to the first, by a central movement.

O     The lines break to the right, and march by that flank along the new position.

#### 15.

B     Is performed by the same movements to the rear, as
O     No. 14 to the front.

---

In all these central movements, it is apprehended much time will not be lost, and the operation would be simplified—if such part of the lines as precede the intersected divisions, are first thrown into the new position, and then that the corresponding movements of the rear part of the lines begin.

*General attentions in the movements of second lines.*

In all these changes of position—the two leading flank battalions must determine, and regulate the movements and distances of the whole, except in the central movements which are directed by the central divisions. Changes on either flank facing either way, and made to either front or rear, as well as innumerable intermediate positions between those above prescribed, are obviously executed by the same methods.

---

Passage of a First Line through the Second.

When in Presence of the Enemy, and that the first line is engaged—the second must be ready to advance

to

to its support; and to give it passage if obliged to retire, without increasing or partaking of its disorder.

When there is reason to apprehend such an *event*, the second line may be prepared for it in the simplest manner possible—by each battalion throwing itself into a double close column of companies behind the two center ones.

Or, the three left companies of the grand divisions of the battalion may double behind their right ones, and in this situation are ready to form in column if necessary. The left divisions of the grenadier and light companies may also double if they are in the line.

Or, the battalions may stand filed from the center of grand divisions to the front, and the flank companies filed from the center.

Preparations of second lines for passage of the first.

Fig. 114.

---

If the first line, or part of it, is pressed and ordered, or forced to *retire*; it will do it as gradually and regularly as it can, fronting (in whole or in part) and firing when possible—It will pass through the openings in the second line. Such parts as present themselves to the heads of the columns, or to the fronts of the companies or files, will incline to the right and left, behind the other parts of the line, and regain their places (which will have been kept open) as soon as they have passed.

The first line will *face* about and form, when about two hundred yards in the rear, while the second line when its front was uncovered will have extended, formed in line, and renewed the action under cover of the fire of the standing divisions, and the particular exertions of the whole artillery redoubled at so critical an instant.

Passage of the first.

---

If the first line passes in disorder, and is so pressed by the enemy, that the second has not time to extend; its columns will decidedly march, and charge whatever is opposed to them—or, march by the flank to attack any body that may have penetrated the

( 166 )

**Movements of the second line when the first passes.**

the intervals, in following the first line—The judgement and decision of commanders of battalions, will in this situation determine much.

These central columns of battalions, may occasionally be formed by both lines, either—to make a partial attack—to follow the enemy—or, the more lightly to make the retreat——The line is very readily formed from them, and on many occasions where movements can only be required in a perpendicular direction to front or rear, it is a very advantageous disposition into which the battalions may be thrown, and to which they should be accustomed.

---

**Passage of the first line by files.**

**Fig. 101.**

But as it may be important to preserve the second line as intire as possible, the most generally received mode of passage, is that by *files* as has been already mentioned; and the rules there laid down for one battalion, are applicable to all those of a line, however strong it may be. The stepping and halting at a determined number of paces from the time of passing the second line, though it will not alone procure an accurate parallel formation is not to be omitted, as it will much facilitate the alignement; especially, if the second line in relieving the first at its post, or in coming to meet it when retiring, joins it in all its battalions at the same instant.

---

**Reforming the first line by a flank.**

**Fig. 101.**

When a line of five or six battalions hath passed in this manner, and that its platoons have halted in column at their determined number of paces, it becomes necessary to give them a more correct dressing before they can wheel up into line—As soon therefore as they are fronted, and that the officers are shifted to their pivot flanks; the commander of the head battalion, will carefully place the pivots of the three first platoons in the true alignement, and will then order the officers of his other platoons to line on the pivot officers before them; and the commander, who

remains

remains with the head platoon as the point d'appui, obferves that this is done correctly—When the firft battalion is thus fteadied, the commander of the fecond places himfelf at the third platoon from the rear of the firft battalion, and fees that his own firft and fecond platoons are well aligned on the pivots before them, and that his others do alfo quickly take up their line. When the fecond battalion is fteadied, the commander of the third proceeds in the fame manner, and fo of the others.

The aligning of the *pivots* muft thus go on gradually, though quickly, from one flank to another, platoon after platoon, and battalion after battalion; any attempt to precipitate it, would only ferve to keep the line in a wavering ftate. Officers muft remain perfectly fteady in their own perfons, on the flanks of their platoons, when they have attained their proper pofition, that the following officers may not be interrupted, but enabled to do the fame thing; the men are fuppofed trained to regulate and aligne themfelves upon their officers.

---

But when an extenfive line hath paffed, and is to be reformed, it may be quicker done in fome cafes, if the *alignement* is taken from the *center* than from one of the wings. The line having paffed, and the platoons being halted and faced in column, the fourth battalion (for example) from the front, is given as the directing one. The platoon pivots of this battalion are firft accurately lined by its commander in the true direction, and become a guide for the whole. The commanders of the adjoining battalions, then placing themfelves each at the third platoon from the front and rear of the directing battalion, aligne the pivots of their own on thofe of that battalion, that are before them—When thefe adjoining battalions are fteadied, the commanders of thofe next them, proceed in the fame manner; and the more the corrected pivots increafe, the eafier does it become to take up the general line; this operation is continued till the whole are juftly arranged.

<sub>Reforming the firft line on a central battalion.</sub>

arranged. Diſtances as well as dreſſing muſt be taken from the fourth battalion.

The battalions that are behind the directing one, will readily arrange their pivot flanks as their diſtances are taken from the front, but the three which are before it will find more difficulty, as they muſt take their diſtances from the rear—To facilitate this, their platoon officers will *face* to the directing battalion, and will then as above, be ſucceſſively directed to take their diſtances from their then front; as ſoon as each has required his true poſition, he will *face* to the right about, and make his platoon join to, and dreſs to him—The line will be ready to form by wheeling up to the pivot flanks.

---

When the firſt line after paſſing throws back a wing.

Fig. 101.

A line which hath paſſed will often throw back a *wing*—In order to occupy a particular poſition—To prevent the enemy's deſigns on that wing—Or at leaſt to make him take a greater detour to effect it—Or that he muſt aligne his own on a height which is occupied and from which he may be flanked.

The line A B. is here ſuppoſed to refuſe its four right battalions after having retired and paſſed as a column with its right in front, which is one of the moſt difficult caſes that can be put—All the battalions of the line to the left of the fourth proceed as is already directed; the right platoon of the fifth battalion is their point of appui, that battalion firſt takes its alignement and gives it as has been directed to the others.

But the four battalions of the right have a more difficult taſk to perform; their appui is the angle where the line begins to break, that is the right flank of the fifth battalion when it has wheeled up into line; their diſtances and alignement are thus both to be taken from behind by the left of their platoons, for the column when halted is formed, as if marched off from the right—This will be the more difficult for the officers to execute, becauſe they muſt conduct the leading flank (the right of their platoons),

and

and therefore cannot time their word *halt* exactly at the inftant when the left is on the alignement.

To remedy this—The commander of the fourth battalion having given nearly to his officers of platoons, the direction which they ought to take, ftops himfelf at the point of appui.—From thence he orders the laft platoon of his battalion to *halt* when he fees its left enter the alignement: the officer who conducted it immediately returns to its left, and facing the commander, is by him correctly placed on the diftant point D, of the alignement; the next platoon is alfo *halted* and dreffed in the alignement by the commander in the fame manner—Thefe two platoons being in order and fronted, their right flanks become a guide on which the heads of the others ftop and do not overpafs.—As the feveral platoons *halt* and *front*, that is, when their right has been aligned with the platoon that is behind them, the officer of each places himfelf on the left flank in his own perfon, faces the point of appui, and is correctly dreffed and lined by the commander of his battalion, as has been already fhown—In this manner do the feveral battalions fucceffively proceed. Fig. 101.

Had the left wing in this cafe been refufed, it would have been much eafier for the officers; for although they muft have equally returned to the left flank of their platoons, yet they would have taken their line and diftance from their natural front.

According to the wing which is to be refufed, fhould the platoons of the line *pafs*.—If the left is to be refufed, they fhould file from their right, fo that the column when halted may ftand as marched off from the right—If the right is to be refufed, the contrary operation fhould take place. What has been already directed for a fingle battalion in crowning a height after paffing to the rear may be equally extended to a whole line. Fig. 101.

In all cafes where it can be done, the countermarch of platoons very much remedies the inconveniences that arife in taking diftances from the rear.

Fig. 103.

Retiring a line formed under the enemy's cannon, and marching upon his flank.

Fig. 103.

The enemy is posted in F G, the line hath advanced to A B, and is exposed to the enemy's cannon—On a more narrow inspection of the ground, his left flank appears the most accessible; it becomes necessary therefore to withdraw the line from the enemy's fire before it can take such a direction as will enable it quickly to gain his flank, and during this operation every means must be used to distract his attention.

In order therefore to place it in such a situation that it may safely prolong its march on a new alignement C E, and not be exposed to the enfilade of cannon—The whole line will break by platoons to the right, and will then execute a kind of gradual wheel by platoons in the direction of B C, and A D, so as to arrive in C D, and from thence prolong its movement upon E, the platoons after wheeling will all face to the right, and in marching in file their heads will bear to the left, and comply with the direction which the first platoon takes—in doing this the following rules must be observed.

Whoever conducts the three leading platoons must take short steps, in order that the rear ones may be enabled to comply with their movements, and every thing depends on their judicious guidance.

Such a gradual direction must be given to the three first platoons, that when they arrive on the line C E. (but not sooner) they may stand perpendicular to it.

All the platoons must follow the direction of the head—That is, their leading flanks must remain nearly aligned upon one another.

The rear platoons may be permitted to take the chord instead of the arch of the circle; and should the first three platoons arrive in the alignement some time before the rear ones, they will not stop but move on at a good pace,

the

the others who may have contracted diftances, can readily regain them in the courfe of following.

In thefe kind of *evolutions* an alignement can be feldom given, becaufe it muft totally depend on the motions which the enemy may make during the movement; the march therefore is generally directed by the leading pivot flanks—When therefore, the general or conductor of the head battalion gives the direction, which the line fhould take, to the two leading platoons; after their officers have placed themfelves on their pivot flanks, the following officers muft carefully cover them and one another, as has been already defcribed—And fhould the commander be obliged to alter the alignement during the march, the following platoons will comply with it, as they arrive at the point where the head changed its direction.

If the corps at A B is formed in two lines—the fecond may with every advantage make this movement, protected by the fituation of the firft, which will keep the enemy in check, and act or follow according to circumftances.

A movement in battalion echellon to the rear, feems peculiarly adapted for withdrawing the line—When that is done, the echellons will front, break into platoons to the right, march till the head of each arrives in the new direction—The leading battalion will halt, till fome of the others are placed behind it, and till it is judged that it may continue its march without occafioning a run or unneceffary ftop in the rear of the column; it will then wheel its leading platoon into the alignement, and be fucceffively followed by the other battalions, till the whole are in one column moving on the enemy's flank.

Fig. 103.

The colums A B which have not yet formed, are in movement nearly parallel to the enemy pofted in C D, and are to continue their march till they take an oblique position E F on one of his flanks.

<small>When a body in column, parallel to the enemy's front, is to continue its march and form obliquely on his flank.

Fig. 104.</small>

Were the columns early to enter on this alignement, the enemy would at once see the intention, and perhaps be able to counteract it; but in order to make him still imagine that he will be approached by a parallel line, and thereby to keep him in his position, the leader of the head platoon of the first line, must during his march, so gradually turn, that when it arrives at E, which is the point d'appui, it may then (and not before) be perpendicular to the line; and the other platoons of that line, having observed their exact distances, and moved on their pivot leaders, will then be all in the line E A—The second line, which is moving in the same manner as the first, will by the gradual turning of its leading platoon, be following out the line G B; and having had more ground to go over, will in proportion have marched faster.

When the heads of the lines are thus at the same instant arrived and halted at their points, the rear platoons of both, that are not in the alignement, immediately file to the left and enter it. The lines are formed, march upon the enemy, who cannot now make a counter movement, and the attack is made from the left by the vanguard, supported by the two lines; which attacks, as has been already observed, should out-flank each other by a battalion at least.

---

Although the INVERSION of all bodies in line, is in general to be avoided; yet there are situations where this rule must be dispensed with, and a quicker formation to a particular front thereby obtained.

<small>Situations where inversions of the line or its parts may be necessary.</small>

The battalion must often *face* to the right about, the more readily to oppose the danger, instead of changing its position by a countermarch—It may even find it necessary to form to a *flank* with its rear rank in front—The column

column with its right in front may arrive on the left of its ground, and be obliged immediately to *form* up and support that point, so that the right of the line will become the left. An army moving to a flank by lines, may be obliged in the quickest manner to form up to the front of its march—Part of a second line, may double up on the extremity of a first line, thereby to out-flank an enemy—Many other situations may be imagined, where opposing the rear rank admits of no choice, and where an inversion of the divisions of the line will gain much time, and becomes absolutely necessary when forming from the point d'appui, and near an enemy.

Troops must therefore be accustomed to such operations; but the application of them requires great method and recollection, otherwise, in such critical situations, confusion is very easily produced, and will ever be attended with the most fatal consequences.

---

At MAGDEBURG—two lines of ten battalions each, marched to the left in two columns of platoons, the interval betwixt the columns was covered by a vanguard of two battalions which marched in front—The enemy having by a shorter route gained the heights on which the columns pointed, it became necessary to form up to the front of the march, in the quickest manner—The two battalions of the van halted, the head of the right column joined the right of the van; the five leading battalions extended to the right in first line; and the five rear battalions also to the right in second line, both in their natural order—The head of the left column joined the left of the van; the divisions closed up to close column, and deployed into line in their natural order, the five leading battalions to the left in first line, and the five succeeding ones in the second line—In this manner, two lines of twelve and of ten battalions were formed, composed of a part of each of the original lines, although no inversion of the battalions in line or

*Manœuvre at Magdeburg.*

*Fig. 88.*

within

within themselves took place—But this operation of *inversion* became necessary, when in order to out-flank and attack the enemy on the left, the five left battalions of the second line marched in column from their left, joined the left of the first line, and then successively extended away to the left, protected by the cavalry—While the five right battalions of the second line, also moved to the left to replace them and support the attack; the five right battalions of the first line remained posted as the wing of appui.

---

Fig. 109.

At MAGDEBURG, the enemy A was posted on an advantageous height; having nine battalions in the first line, two battalions in the second line, and an advanced corps of three battalions about four hundred yards in front of the right, which was covered by a village, while the left extended into the open ground, but protected by the whole cavalry of ten squadrons.

The attacking corps, having from columns of march formed in two lines of nine and ten battalions, with an advanced guard of three battalions, marched from the left, each line in three echellons B, drove back the enemy's advanced corps, and seemed to threaten an attack on his right: but finding it too strongly posted, in an instant the disposition was changed, and a determination taken to act on the other flank—The whole corps was ordered to march by sections to the right: the vanguard together with the six left battalions of the first line, formed in a waving manner along a favourable height C, and remained fixed as the wing of appui, and to keep the enemy's right in check—The three battalions of the right, together with the whole second line, continued their march about five hundred yards farther, circling towards the front, and then halted in three parallel columns D (the heads pointing towards the enemy's left); the left one composed of three battalions, the middle one of four battalions, and the right one of six battalions.

The cavalry of ten squadrons which covered this march, was now ordered to move on that of the enemy, and drive them back

to

to their line from which they had advanced—The three columns of the right having doubled up to platoons, moved on under cover of a rising ground and of their own cavalry, till abreast of the enemy's left; and having in the course of the march preserved exact distances, they at once by wheeling up to the left, formed in three lines E; the second out-flanking the first to the right by one battalion, and the third out-flanking the second, in the same manner, by two battalions—The lines having thus formed within about five hundred yards, immediately marched on, and at the distance of about one hundred and twenty paces, began to engage the enemy with musketry; who on seeing that his left would be attacked, had marched the three battalions of his vanguard, and the two battalions of his second line, and was forming them on the left of the whole to prevent being out-flanked at F.

Fig. 109.

The first attack of three battalions after firing some time, and being repulsed, retired by platoons by files through the other two lines—The second attack of four battalions moved up, engaged, and after some time, retired in the same manner. There now remained the third line of six battalions to cover the retreat, which after firing two or three rounds on the enemy, who began to advance, commenced a retreat *en echiquier* by the alternate battalions, and continued it, supported by the cavalry G for about five hundred yards, till it had arrived under the descent; from thence, as the enemy would not quit their advantageous position to pursue farther, it broke into columns, and joined the line.

The two first bodies which had attacked and were repulsed, being supposed to have suffered too much to form sooner, had continued to retire in platoons by files, till they arrived and joined the corps C which had remained posted on the left, as the point d'appui, and on which the whole assembled.

## Of the RETREAT.

<small>General method of retreating.

Fig. 110.</small>

ALL manœuvres of a line in RETIRING, are infinitely more difficult than those in advancing—They must be more or less performed by chequered movements, one body by its numbers or position, facing, and protecting the retreat of another; and if the enemy presses hard, the whole must probably front in time and await him.

As the ground narrows, different parts of the corps must double—Mouths of *defiles* and advantageous posts must be possessed in time; and by degrees the different bodies must diminish, and throw themselves into column of march. The general principles of movement here apply, adapted to circumstances and situation.

---

If a line with *reserves* finds it necessary to retreat in face of an enemy—

<small>Retreat of a line with reserves.

Fig. 110.'</small>

The alternate battalions and reserves, will retire two or three hundred yards and then front: the other battalions will then retire; and when they join the first, the reserves also will march and front at like distances, the reserves always leading the retreat.

This will continue till it is proper or safe to break into column of march—The cannon and skirmishers of the whole, covering the front of the retreating line.

The CHEQUERED RETREAT, or that by the alternate battalions of a line going to the rear, while the others remain halted and cover them, is the quickest manner of refusing a part of any corps to the enemy, and at the same time protecting its movement, as long as it continues to be made nearly parallel to the first position.

*The chequered retreat in line.*

*Fig. 112.*

If ten battalions are in line, the five even ones counting from the right, will go to the right about, retire in line about two hundred and fifty paces, and then front; having carefully preserved their intervals—The two outward battalions only of the retiring divisions, will each when it faces about, form a flank of its outward platoon.

As soon as the second division begins its retreat, all the battalions of the first division, will immediately throw back their wing platoons, and thereby when necessary, procure a cross fire in the intervals and along the front.

When the second division fronts, the first is ordered to retire through the intervals, and to form at an equal distance in the rear; and in order the better to cover those intervals, the flanks of the battalions (except the two outward ones) will move up into line, when the division fronts.

As soon as the first division arrives near the second; that second begins to fire by platoons standing, in the same manner as the first hath already done. The wing platoons of all the second division battalions, place themselves in flank, as soon as the first hath passed them; and when they themselves have retired, and fronted at their proper distance, they will move up into line as above, and wait the retreat of the first division.

During the retreat, should favourable heights or situations present themselves to either of the divisions—they should be for the time occupied by the most contiguous battalions, who will halt or incline as is necessary, without scrupulously adhering in that case to the alignement or intervals; and any battalions that may

A a happen

happen to possess an advanced height, should throw their wings back, and aligne them on their neighbouring ones, that they may be flanked by such battalions.

When such detached battalions are in their turn to retire—they must not pass in file through any other battalion that may happen to be in their rear, but must incline as is necessary to regain their proper interval, and to resume their proper place in the division.

---

**Refusing a wing in retreat.**

**Fig. 112.**

A *wing* may be refused during this retreat, in the same manner as has been directed when the lines pass through each other—When the advanced wing halts and fronts, the wing that is to be refused, moves on; and acquiring from the gradual turn of its colours, an oblique direction, halts when proper, and fronts in its new intended alignement. The first division, in the course of its retreat, will gradually conform to the situation which the second has given, move through its intervals, and cover them exactly when it fronts.

---

**Chequered retreat of two lines.**

**Fig. 113.**

*Two* full *lines* will make their retreat, in the same manner as has been above prescribed for one—If the distance between the lines is three hundred paces, each will give to its second division one hundred and fifty paces for its retreat, and thus divide the distance.

When the second division of each line hath retired and fronted—the first division of the first line, will retire through the intervals of its own second division; it will then, when it arrives at the first division of the second line, *pass* by *files* through the battalions of that division; and in the same manner will it pass through the intervals of the next division; and when one hundred and fifty paces in the rear of the whole, the platoons will halt, front, and wheel up into line, as has been before directed.

The

The second division of the first line having prepared its flanks, begins to retire, as soon as its own first division hath passed the first of the second line—That second division will march through the intervals of that first; it will then proceed, and pass by *files* through the battalions of the second division of the second line; it will continue in that shape, and pass through the intervals of its own first division, which is by this time reformed; and when at its one hundred and fifty paces in the rear of the whole, the platoons will halt, front, and wheel up into line—The divisions of the second line proceed in their turn, exactly in the same manner.

The following general rules must be observed.

The battalions of the division nearest the enemy, will form flanks as soon as there is nothing in their front to cover them; but the other divisions will have no flanks except to the outward battalion of each.

The battalions always pass by their proper intervals— and it is a rule in retiring, that the left of each shall always pass the right of the neighbouring one.

<small>General rules observed in the retreat.</small>

Whatever advantages the ground offers, are to be seized, without being too critically tied down to intervals, or to the determined distance of each retreat.

The division next the enemy, must always pass in front through the intervals of the division immediately behind it; and any battalion that finds it necessary, must incline for that purpose.

The retiring divisions must march nimbly, and take no more time than what is necessary to avoid confusion.

The division nearest the enemy, fires by platoons standing—The flanks of its battalions only fire when the enemy attempts to push through the intervals; when that division retires, it fires on skirmishers by single men, and on small bodies by files, without altering its ordinary march: but should any of its battalions be obliged to fire by platoons retreating, a shorter step must then be taken; and should the enemy threaten to enter at any of its intervals; besides the fire of its flanks, such platoons of the line behind it, as can with safety, must give it support.

---

There are various opinions as to the comparative advantages of CAVALRY and INFANTRY, when opposed to each other; and as to the manner in which the latter should resist and repel the attack of the former.

Experience has often shown that our thin lines of infantry are unequal to the situation, and when once that order is broken, on which a soldier has been habituated to repose his confidence and security, it is in vain to appeal to the reason or spirit of the individual; resistance is generally given up, panic prevails, and flight however unavailing universally takes place.

*General circumstances where infantry are to oppose cavalry.*

Fire alone certainly ought not to stop the progress of a determined cavalry, and it is hardly credible how few men and horses are at the instant brought to the ground, by the most steady and well directed fire; therefore it seems eligible in some situations to prepare openings, towards which the cavalry will naturally swerve, and through which perhaps the whole will find their way.

But undoubtedly, there is much danger in allowing the line to be pierced, or in altering a disposition at the instant of being threatned by cavalry: and therefore in *line* of *battle*, where the flanks of the army are covered, where the getting round them would be a considerable and critical operation, and where the uniform front is to be maintained, the attack of the cavalry is at any rate to be opposed by steadiness, supporting corps, and a heavy constant well directed fire of musketry and artillery—Notwithstanding these, should a part of the enemy break through the line, it is an event that ought by all to be expected, but not without its remedy—When the troops are thus prepared, they will be the lefs furprifed to fee cavalry in their rear, who cannot long remain to advantage between the lines, under a fire in all directions (if the infantry are fteady) and who alfo are liable to be attacked when in diforder by the fupporting cavalry.

---

There are fituations, in which infantry are expofed to the attacks of cavalry, when they cannot nor need not remain in extended line, and when they ought to take fome other fhape, in order to give fuccefsful oppofition—This will happen, when a corps not very confiderable, cannot prevent the enemy's cavalry from getting on their flanks or rear, either when the retreat is made in *line* of *battle*, or—when in *line* of *march*, they are moving without baggage, or with baggage. *Situations in which infantry are expofed to the attacks of cavalry.*

When retiring in *line*, that the attack is impending, and that it is judged proper to form on a greater depth—As the movements of cavalry are rapid, thofe of the infantry ought to be fimple and quickly executed; the openings ought to be confiderable, and the front of the troops oppofed not too great, that the cavalry may the more eafily avoid them if they are fo inclined. The battalions may therefore form clofe columns of platoons, behind the right of each grand divifion; and the left platoons of the grenadier and light companies will double. *When retiring in line.*

The enemy will be received by the front platoon kneeling and prefenting their bayonets; the two next ranks of the fecond platoon will keep up a file firing. If neceffary, half the column will in the fame manner, face to act to the rear—If attacked on the flank, the right or left fubdivifions will face, the two front ranks kneel and prefent bayonets, the two next ranks fire. When the enemy is repulfed, the line is readily formed. *Fig. 114.*

Every

Every effort of fire must be made to prevent the enemy entering the line; and whether there is one or two lines, if these columns by a degree of wheel take a diagonal position, a cross fire in every direction is obtained, and the instant before being attacked, any particular column may readily wheel up to present its full front.

*When retiring in line.*

*Fig. 115.*

The position of such columns or battalions, will much depend on the judgement of particular commanding officers, in such a critical instant. But the great object is to preserve compactness and order; and to resist the impression which the first onset of the cavalry must necessarily occasion.

When *marching* without baggage, which will be the case when infantry retire after an unsuccessful action; the columns of retreat will be the general one, formed by companies—each column may consist of two or three battalions at most—the number of columns increase the relative strength of the retreat—all embarrassment of baggage is supposed to be previously sent off—every country must soon present points, which being occupied, will give security to the retreat, and alter the nature of it.

*When marching without baggage.*

If the columns are obliged to halt, they will be closed up; and to whatever front the enemy threatens an attack, they will present three ranks kneeling, with their bayonets pointed and fire reserved; and two ranks firing by files or by ranks.

When enabled to resume the march, the divisions may open out in some small degree to accelerate the movement.

When *marching* with baggage, or where a *convoy* is to be protected, a particular disposition must be made, arising from circumstances of situation, &c. At all events, some solid and considerable body should be formed, which ought apparently to resist the efforts of cavalry, and on which the more inconsiderable bodies may rally if broken: the nature of the country, as wood, rivers, &c. ought to give protection on one flank at least—the enemy must be opposed by movements and positions—if he is once enabled

abled to break the line of march, the greatest confusion must ensue.

---

Where bodies of one, two, or three battalions are to retreat, they are better thrown into a rectangular form than into that of a perfect square. The retreating front should be of two or three companies at most—the sides should march by divisions and be six deep when fronted, or at least from the internal arrangement of the figure, that number should be ready to be opposed to wherever the enemy presents himself. {Column of retreat for several battalions. Fig. 116.}

It is therefore from circumstances, and from the flexibility of the military order, that the commander may in an instant determine and direct into what shape he will throw the body which he conducts.

On all these occasions, where columns or solid bodies are opposed to cavalry, it cannot be imagined that they are at the same time liable to the attacks of a considerable infantry or artillery; such situation would be critical indeed, and from which nothing but the most determined resistance could extricate them.

---

The SQUARE or OBLONG, is a shape which infantry have at all times taken, when obliged in open ground to march in the face of cavalry. Though the mode of placing one battalion in this manner, may be determined; yet the various formations of which a greater number are susceptible, cannot be ascertained: they depend on ground—the position of the troops—the movements of the enemy, &c. and must be made in consequence of the local orders of the general.

---

At BERLIN, six battalions retiring in line in open ground upon a strong position, about a mile distant in the rear of its left, and
which

which was already occupied by three detached battalions, was obliged on the approach of cavalry on its right to form an oblong, and in that shape to make its retreat.

*Manœuvre in the square at Berlin. 1785.*

*Fig. 102.*

The two left battalions advanced about twenty paces, halted, and faced their outward flank platoons to the flanks—The right battalion went to the right about, wheeled back upon its left, and again fronted, covering the right of the line—The third battalion from the left wheeled backwards by platoons on their right, and then marched in column, the left in front to form the left face of the square—The second battalion from the right filed from the right gradually to the rear, turned to the right, formed into column of sections; during its march, was followed in the same manner, and in the same column by the third battalion, moved on to form the rear face of the square, and was protected on its march by the front face which was already formed, and by the right battalion, which going to the right about, by degrees closed into the right of the front-face, in proportion as those battalions which were filing into column made place for it, and finally became the right face of the square.

The square, or rather oblong was now formed, and the angles were broke by each face throwing its flank platoons inwards.

The square now marched to the left by its left face—the square halted and fired to its several faces—the square marched by its rear face, and when about three hundred yards distant from the several protecting detachments, it halted, and fronted.

The side faces now wheeled into column, and continued their march into the defile; the front face retired on the rear face (which then fronted to the enemy) and when near it, faced about, fired, then passed it by files, and again formed near the mouth of the defile.

The rear face which had thrown back its flanks now retired, past through the front face, protected by the fire of the posted detachments, and marched in column into the defile—The front face followed it, and the rear of the whole was closed by the several detachments gradually retiring, possessing the several commanding

manding heights, and lessening by degrees till they occupied the defile.

The whole was well executed—the sides of the square wheeled, and marched by sections, not by files—the forming of the square was well protected, and the reduction of it also well imagined.

## Of the DEFILE.

The PASSAGE of the DEFILE, will either be in front or rear from column—or, in front or rear from line.

When the column marching to the front, arrives at the head of the *defile*—according to the nature of it, the companies, the platoons, the subdivisions, will gradually, and as they pass the narrow ground, double behind the right, if the column has the right in front; or behind the left, if the left is in front. Subdivisions if necessary, will also file from the same leading flank, and distances will be closed up.

*Passage of the defile to the front of the column.*

*Fig. 111.*

As the ground opens, the file will form subdivision, platoon, company, column, and proceed in march, or form in line as ordered.

When the column which has been retiring by *its* rear ranks leading, arrives at the mouth of the *defile*, it will halt and front;

**Paſſage of the defile to the rear of the column.**

every previous meaſure will have been taken for its entering the defile with ſafety—The rear diviſions will break ſucceſſively as already preſcribed, the rear ranks leading. When the defile is paſſed, the diviſions will double up to the marching front, and reſume column; which if the line is to be formed, will halt, front, and extend into line.

If in retiring the column is not too much preſſed, inſtead of marching by the rear rank leading, and ſuppoſing it originally formed with the right in front; it will countermarch by diviſions, and become a column to the rear with the left in front, in which ſituation it will purſue its march and paſs the *defile*. When neceſſary to face, the diviſions will again countermarch.

**When the defile is before one of the flanks of the line.**

When the line is to paſs the defile to the front, and that ſuch defile is near to one of its flanks, the line will break into column from ſuch flank, and paſs on ſuch front as it will permit.

Fig. 105.

**When the defile is before the center of the line.**

When the defile is nearer the center, the line will break inwards to that point and paſs in double column, while there are troops from each flank to compoſe it—If there is a ſurplus on one flank of the line, it will then follow, but on as great a front as that of the double column.

Fig. 105.

When the line is to paſs a defile in its rear, it will, according to circumſtances, march from either flank in column, or in double column from the center.

**When the defile is behind a flank of the line.**

If the defile is behind one flank, the line may break from the other, and the column march along the rear, and under the protection of the parts of the line which remain formed, till it is neceſſary that they ſhould take their rank in column.

Fig. 106.

If the defile is at, or towards the center—equal portions of the line on each ſide of the point neareſt the defile will be marked, the line will march off from the outward flanks of ſuch portions,

meet

meet in the rear of the center, and proceed in double column to pass the defile. In the mean time, whatever surplus remained on one wing, will have marched by its inward flank towards the central point, and on a front equal to that of the double column the rear of which it will follow—or, such surplus may be first placed in front or rear of the central point, considered as a separate line, and from that situation follow, or lead in double column through the defile.

When the defile is behind the center of the line.

Fig. 106.

---

The several modes of diminishing or increasing the front of the column, as circumstances require, and the consequent extension into line are obvious, and have already been prescribed.

It can seldom be necessary to begin by filing from one or both flanks of the line in passing to the rear, or from the center, or one flank of the line in passing to the front: but the march in divisions should be preserved as long as possible.

The defile ought to be filled but without crowding, and passed at as quick a pace as order will allow.

When diminishing from column and passing the defile, the front must move on—When doubling up to column, the head must halt, or must be accurately and attentively conducted, that the rear may not be too much hurried.

The line on quitting the defile, should not be formed too near the mouth of it, to prevent any improper stop.

Both mouths of a defile ought to be possessed in force before it is entered by the troops, and previous measures taken for passing it in safety to the front, and extending into line; or, for the protection of the last bodies of the line that enter it when passing to the rear.

Attentions in passing a defile.

When the divisions of a battalion or column break into sections or subdivisions, in order to reduce the front of the march, it will always be done by doubling behind, regularly from the one to the other flank of each; and never by marching off from the center of such divisions, which would thereby unnecessarily break in upon the universal principle, that every part of the body when in column, should uninterruptedly follow from front to rear.

<small>Method of reducing the front of the march.</small>

Each division breaks when it arrives at the spot (and not before) where the first division has broke off, that is, at the mouth of the defile. Arms must be carried steady—ranks as well as sections must be closed, and an equal march preserved—the doubling sections must not wheel but incline—the sections must be from four to six file in front, otherwise the battalion will lengthen out, and cannot march on the same extent of ground it takes up when in line.

---

The sections form up when necessary by inclining, and as they arrive at the point where the first forms up, which will moderate its step to prevent a run in the rear.

<small>Increasing the front of the march.</small>

In marching in open ground if distances are preserved, the front of the column may be increased or diminished, by the sections inclining up, or falling back all at the same time.

---

The line will on many occasions of service at once break into column of sections, as a convenient front on which to pursue the march; and if those sections are of six file each, the rear ranks can be sufficiently loosened and march with great convenience, and without increasing the distances of divisions. In this manner, the PRUSSIAN line always breaks when it is to occupy any considerable height, and the column winds along it in this order, as more flexible than that by platoons, or on a greater front.

<small>Marching by column of sections.</small>

---

<small>Fig. 107.</small>

If a line is to cross a bridge in front—such line would file from the center of whatever battalion is nearest to it, and form to that battalion on the other side.

If

If a bridge is to be crossed in the rear—it will be done by filing from both flanks of the line or lines succeffively, while the center protects the movement.

Fig. 108.

## Of the Column of March.

The Column of March—is the foundation and mean of all movements whatever, and of all tranfitions from one pofition to another.

Whether it is compofed of a *file*, a *divifion*, or a greater *body* in front, the principles remain the fame. That it fhould never occupy a greater extent in marching, than is equal to its front when in order of battle—That it fhould march on as great a front (as far as that of a company or grand divifion, for a greater would be fatiguing, if a confiderable fpace was to be gone over) as the ground will admit of—That as many columns fhould in general be formed, as the ground and routes will allow—And that all thefe fhould be regulated as to diftances and direction, by fome one certain column.

Principles of the column of march.

Thefe objects fhould be held in view as facred, and not to be deviated from without the moft urgent neceffity—They are enforced by the practice of all fervices but our own—Indifpenfable

in

in the operations and conduct of great bodies, they are therefore to be strictly observed in the movements of small ones, that their multiplied errors may not affect the grand machine.

---

**Columns always to march by divisions.**

At *home*, our battalions are too little accustomed to march and act with others, as part of a large body. On the common roads we are obliged to give way to every obstacle, and fall into file; so that the necessity and importance of division marching, is not sufficiently impressed on the minds, or attended to by British officers. Yet it is only in that order, that the battalion should at any time perform its marches; that the columns of an army should be permitted to move; that an enemy should be approached; and that safety can be insured to the troops in their transitions from one point to another.

**File marching never to be practised in column of march.**

Therefore, the marching of great bodies by *files* (as an extension of such a column is unavoidable) though much practised in the British service, must be looked upon as a vicious and unmilitary mode, only to be applied where the difficulty of ground will allow of no greater front; and division marching ought on all occasions to be used by all bodies great or small—Where any considerable distance is to be gone over by the division, battalion, or line in front; it will be permitted to the rear ranks during the march, to open two or three feet, and to close up again whenever a halt is made.

---

**Marching in inclosed countries.**

Where a country is so inclosed, woody, or the routes so bad and narrow that it is absolutely necessary to march in *file*, or on a small front, there is no remedy for the delay in forming; and man may be obliged to come up after man: but these circumstances, which should be regarded as exceptions from the primary and desired order of march on a greater front; should only enforce the great principle of preventing improper distances and of getting out of so weak a situation as soon as the nature of the ground will allow the front of the march to be increased, or the line, or part of it to be formed.

In such difficult situations, the number of bodies the line can break into will be few, the front they march on will be small, the rear of course extended: but still the same great objects of regularity and precision are to be held in view, and are to be attained by the same methods, although more time is required.

---

In common *route* marching, the same regularity of step cannot be required, as is necessary in the operations of manœuvre—The battalion or column may be carried on at a natural pace of two miles and a half per hour; the attention of the men may be relaxed, and the ranks and files loosened, so as to move with the greater conveniency, but never confounded; the proper distances of divisions must never be increased, and the proper flank men and officers remain answerable for them—No diminution of the front, on any account but by order of the leader of the column; bridges, and short defiles to be filled and passed with additional quickness; the head to move in such a manner as neither to hurry nor stop the rear of the column—When the column arrives near its object of formation, or manœuvre, the strictest attention of officers and men is to be resumed, and each individual is to be at his post.

*Attention in common route marching.*

---

It is always time well employed to halt the head of a column, and enlarge an opening, or repair a very bad step in the road, rather than to diminish the front or lengthen out the line of march.

No individual is to presume to march on a less front than what the leader of the column directs, and all doublings must therefore come from the head only. The closeness of the march on all occasions is a point of the highest consequence, and it is a most meritorious service in any officer to prevent all unnecessary doublings, or to correct them as soon as made: and on all occasions to march on the greatest front, the roads or overtures will allow, although the regiment or divisions before him may be marching on a narrow front.

*The rear of a column never to lengthen out.*

At

**Attentions on every increase or diminution of the front of the column.**

At all points of increasing, or diminishing the front of the march, an intelligent officer per battalion, or brigade, should be stationed to see that it is performed with celerity, and the commandant should have constant reports and inspections made, that the column is moving with proper regularity, he should have officers in advance to apprize him of difficulties to be avoided, and should himself apply every proper means to obviate such as may occur on the march.

**Attentions of commanding officers during the march.**

At all times when commanding officers see that there are likely to be impediments from the nature of the ground, to the movement, or march of their regiments—They should detach officers in advance, to reconnoitre, and point out the means, and openings by which such obstacles are to be passed; and at no time are such helps so necessary, as when regiments are acting in line, in broken ground, and when their movements are combined with those of others.

**Necessary attention of march.**

All overtures made for the march of a column should be sufficient for the greatest front on which it is to march, and should be of the same width, otherwise each smaller one becomes a defilè.

The distance of columns from each other during the march depends on the object of that march and on the nearness of the enemy.

The more columns in which an army marches, the less extent in depth will it take up; the less frequent will be its halts, and the more speedily can it form in order of battle to the front.

---

**Combination necessary in marching.**

On the *combinations* of march made by the general, and on the execution of these by the troops, and by the leaders of the several component parts of the army does the success of every military operation, or enterprize depend.

A perfect knowledge of the country, of the particular distances to be gone over, and of the probable obstructions or delays which

each

each column may separately meet with, determines the hour of its departure, and that of its arrival at the given rendevous.

To fulfil these intentions of the chief requires in critical important situations, every concurrent exertion of the subordinate officer—The *theory* of marches may be known in all services, but the just *practice* takes place in few; and innumerable are the instances of the best concerted dispositions failing, from a want of that punctuality of execution, which every general must trust to, and has a right to expect from the leaders of his columns.

---

The composition of the columns of an army, must always depend on the nature of the country and the objects of the movement.

*Composition of column.*

*Marches* made parallel to the front of the enemy, will generally be performed by the lines on which the army is encamped; each marching to the flank at a small distance from one another, and occupying when in march the same extent of ground, as when formed in line. By this simple movement, has the flank of an enemy often been gained by troops accustomed to diligence and precision in the execution of their marches; nor (where the ground allows of it) is there any manœuvre of a great army more important, or that can be more securely or effectually practiced against an enemy inaccurate and inferior in discipline, who in attempting the counter movement, is generally thrown into confusion—The battles of *Prague*, of *Rosbach*, of *Lissa*, among others are examples of this truth, and of the superior movements of PRUSSIAN troops.

*Marches by lines to the flanks.*

---

*Marches* made perpendicular to the front of the enemy either advancing or retiring, will be covered by strong van or rear guards—The columns will be formed of divisions of the army, and each generally composed both of cavalry and infantry, the nature of the country will determine which arm precedes.

*Marches to front or rear.*

---

During a *march* to the *front*, the separation of the heads of the columns, must often unavoidably be considerable: but when near the enemy, they muſt be ſo regulated, and directed, as to be able to occupy the intermediate ſpaces, if required to form in line—Some one column muſt determine the relative ſituation of the others, and from time to time, new points of rendevous will be given: the diviſions muſt be more cloſed up than in the march to the flank, and in proportion as they approach the enemy, muſt exactneſs and attention increaſe—The general in conſequence of the obſervations he has made, will determine on his diſpoſition; the columns which are now probably halted and collected will be ſubdivided, and multiplied; each body will be directed on its point of formation; and the component parts of each will, in due time, diſengage from the general column and form in line.—The ſafety of *marches* to the *rear*, muſt depend on particular diſpoſitions, on ſtrong covering rear guards, and on the judicious choice of ſuch poſts as will check the purſuit of the enemy.

---

In theſe marches to front or rear, the diviſions of the ſecond line generally accompany thoſe of the firſt, and all their formations are relative thereto—The heavy artillery and carriages of an army, form a particular object of every march, and muſt be directed according to the circumſtances of the day—The ſafety of the march by the arrangement of detachments and poſts, to cover the front, rear or flanks of the columns, depends alſo on many local and temporary reaſons, but are an eſſential part of the general diſpoſition.

---

Column of march formed from line.

Fig. 119.

The *column* of *march* will generally be formed from line by the ſeveral bodies which are to compoſe it, wheeling to the one or other flank—The leading diviſion then moves on in the given direction; and the reſt follow, preſerving ſuch diſtances as are ordered, and which depend on the objects of the march.

To front or rear.

If the probable formation of the line is to be to the front, or rear of the march—The diviſions will be ordered to follow, at half,

or

or quarter intervals, in order to contract the length of the column consistent with the views of the commander.

Fig. 121.

But there is no possible case where a battalion or greater body in march, ought ever to occupy a greater extent from front to rear of the column, than it does from right to left when formed in line.

If the probable formation of the line is to the flank of the march, the divisions will always preserve the distances they wheeled at equal to their front; so that the line may at any time be formed to the flank, by the divisions halting, and each wheeling the quarter circle.

To the flank.

Fig. 120.

When a *large body* marches from its camp, or position in several columns to front or rear, those columns may be formed nearly close (where circumstances and the ground allow of it) in front and rear of such divisions, as are opposite the openings that conduct to the new *position*; and they will afterwards take their marching distances as ordered.

Assembling columns preparatory to the march.

Fig. 123.

If the impediments of ground, hinder them from forming the close column to the front—They will close in, in the prolongation of the line, to the leading division of the column; and then follow it along its prescribed tract, and at such marching distances as are ordered.

Fig. 122.

If the *march* is to the flank of the camp, or position, the bodies that compose each column, will close into the division that is to lead, and then follow it in whatever direction is allotted to the column, and at whatever distances are ordered.

Fig. 124.

The following examples will show the general movements of an ARMY from one position to another, and the different manners of forming in line from different columns of march.

**March and attack of a corps posted.**

**Fig. 117.**

The enemy A is posted with five battalions and six squadrons, his left at a morass, his right extending along a rivulet.

The column of eight battalions, and three squadrons advances towards the center of the enemy, and halts in battalions and squadrons behind a wood: three other squadrons to the left, form the advanced guard.

The position C is to be taken, and the attack made on the enemy's left—the battalions in rear of the first and fifth close up to them—the eighth battalion remains as a reserve behind the wood; the other seven battalions and three squadrons march by divisions to the right, and take the position B. They will then cross the rivulet together; form in three lines C, make the attack on the enemy, and be favoured by the other three squadrons and the eighth battalion, whose motions will correspond with that of the attack.

The enemy when obliged to quit the position A, throws back his right, and takes post D on the height, from whence he makes his retreat.

---

**March of an army in different situations.**

**Fig. 118.**

An ARMY at A, marches to the right to B in four columns, and by lines as encamped, viz. two columns of infantry, one of cavalry, one of artillery, and baggage.

From B it advances in four columns to C, marching by divisions of the army; each column composed of infantry, cavalry, artillery, baggage, and in the same order; the several columns preceded by a strong advanced guard.

From C it retires to its first position A, marching in four columns by divisions of the army; each composed of baggage, heavy artillery, cavalry, infantry—The second line of infantry precedes the first, and a strong rear guard covers the whole.

The particular composition of the columns will vary, according to the nature of the country, and other circumstances.

An ARMY consisting of thirty-nine battalions and thirty-six squadrons, encamped at A in two lines of infantry; and the cavalry in a third line, marches in four columns to attack the enemy—Each column is composed of infantry of both lines, and also of a part of the cavalry—The infantry precedes—All the columns march from the right.

*March and formation of an army previous to the attack.*

*Fig. 125.*

The advanced guard, consisting of ten squadrons and four battalions, precedes the center columns, and is halted at C, when the heads of the columns arrive at a b c d.

After reconnoitering the position of the enemy, the commander resolves to attack their left; six squadrons of the advanced guard march to D; four squadrons and four battalions march to E.

The columns then close up to half or quarter distance, and are carried on to f g h i.

New points are then given at k l m n, to which the infantry of the respective columns march; and the cavalry of the three right columns are ordered to form two additional columns on the right, viz. that of O, moves up to P; that of N, followed by M, moves up to O; the cavalry of L accompanying its infantry—The heads of the several columns are then carried to the points q r s t u w, where they form in order of battle—The two columns of cavalry in the prolongation of their line; the left and the two right columns of infantry by deploying from close column; the other column of infantry by flank marching to the left.

The attack is then made from the right, either in line or in echellon; the reserve E supporting the attack, or if repulsed covering the retreat—The infantry H moves on, as the attack of the right succeeds—The cannon which preceded or followed the column, is placed in front of the line.

---

In order to facilitate the MOVEMENTS of great ARMIES, and to enforce their discipline, it is necessary to organize,

divide,

**Necessity of regulation for the marches of an army.**

divide, subdivide them; to establish such general regulations, as may prevent the repetition of a tedious but essential detail, at times when action and exertion are required; and particularly to ascertain the duties and attentions of individuals, in every situation of march; so that the most concise orders may suffice to put the army in motion, and to place it at all times ready to execute with exactness and alacrity, such intentions of its commander, as may arise from the circumstances of the moment.

Those *rules* therefore, that more immediately regard the marches of an army, are here exemplified in the practice and ORDERS given for the marches and movements of the BRITISH and FRENCH armies in *Germany*, during the war of 1760. The sameness of principle which directed the conduct of Prince FERDINAND and Marechal BROGLIE, is sufficient to sanctify their justness; and by both generals, the main armies, the reserves, the detached corps, light troops, and van-guards, were moved and applied according to circumstances, by the same springs of action, and as parts of one and the same great machine.

---

### Marechal de Broglie.

" Besides the accustomed distinctions of battalions, regiments, brigades, wings and lines; the ARMY will during the campaign, be formed in six divisions; four of infantry, and two of cavalry.

" The two lines of infantry will compose the four divisions; each of a fourth part of each line.

### Prince Ferdinand.

The ALLIED ARMY, from the nature of its composition did not admit of the same regular and equal divisions as the French army; from time to time, they however existed.

It was generally understood as forming six divisions—The cavalry of each wing one—and the infantry of the four nations, British, Hanoverian, Hessian, Brunswick, one each.

The

## Prince Ferdinand.

The generals commanding nations, brigades, &c. were charged with their respective discipline and police.

There was for the day, one lieutenant general, one major general for each wing of the army; one field officer for each wing of cavalry, one field officer for the infantry picquets of each nation.

A brigade of light artillery, consisting of twelve six pounders attached to the infantry of each of the four nations, and always marched in column and encamped with it.

---

I.

" The army will march either by divisions or lines; in the first case, it will form seven columns; in the second, four columns.

" When by divisions, the cavalry form the columns of the right and left, the heavy artillery that of the center, and the infantry the other four columns; each attended by a light brigade of twelve six pounders.

When by lines—the first column will be composed of infantry of the first line with twenty-four six pounders; the second of the infantry of the second line with twenty-four six pounders.

The third of the heavy artillery; the fourth of the whole cavalry.

The twelve six pounders attached to each of the four nations, are to march between the first and second brigades of that nation.

In the rear of the battalion which closes the column, are to follow the ammunition

## Marechal de Broglie.

" The right wing of cavalry will form one division, the left wing of cavalry will form the other.

" A brigade of infantry will be also named to cover the flanks of each wing of cavalry.

Each division will be commanded by its own general officers, appointed for the campaign; nor will those of another division ever interfere.

The commanding lieutenant general of each division has charge of the discipline, police, marches, communications, &c. and to him all the inferior general officers report.

Each division will have its general officer or brigadier of the day to attend to the above-mentioned objects, and report to its lieutenant general.

Detachments will be furnished as much as possible by divisions.

---

" The army will generally march in six columns, or four columns.

When the army marches in six columns, each column will be formed by a division, viz. four of infantry, two of cavalry

The right division of infantry will be called the first division; the others in succession to the left will be the second, third, fourth.

The

## Marechal de Broglie.

The cavalry will be named, right or left divisions, or wings.

When the army marches in four columns, the first line of the right wing of cavalry will march with the first division of infantry, and the second line with the second division.

The first line of the left wing of cavalry will march with the fourth division of infantry, and the second line with the third division.

The nature of the country will determine whether the cavalry ought to lead or follow the columns of infantry.

A division of artillery will always follow each division of infantry.

The flank brigades of infantry will be ordered in front or rear of the cavalry, and under the command of the general officer of the wing to which they are attached.

Thus the general order of march will be—troops, artillery, light carriages, heavy ones.

The artillery of the park will always follow the carriages of such column, as marches by the best route, and which will be prescribed.

———

The march of the army will never be announced in orders.

## Prince Ferdinand.

waggons of all the regiments, according to the order of march of the battalions and brigades—And then the ammunition waggons of the six pounders attached to the column.

The columns of cavalry having neither cannon nor ammunition waggons, their equipage will immediately follow the squadron that closes the column in the order prescribed.

The equipage belonging to the columns formed by the heavy artillery, will follow the last ammunition waggon, and the other carriages of the artillery

2.

The order of march will be given in the following manner.

The army will march the ——— exactly at — o'clock.

The order or notice of march will be sent to each nation—it must be received by the eldest general officer, colonel, or lieutenant colonel present, and executed in the following manner.

Half an hour before the time fixed for the march, the general commanding each nation, will give the signal for that purpose; the tents must be struck and baggage loaded immediately—Each brigade must be formed at the time appointed, and the baggage remain in the rear ready to follow.

The commanding officer of each brigade will order a subaltern to conduct the carriages with regularity.

Besides the notice of march, the eldest general, colonel or lieutenant colonel, of each nation present in camp, will receive a sealed order, on the outside of which will be marked the

## Prince Ferdinand.

the time for opening it. This order will contain the difpofition of march.

### 3.

When the army is to march in feven columns—The brigades which will be ready formed in order of battle, muft begin their march at the moment prefcribed to form in column in rear of each other, the infantry by battalion, the cavalry by regiments—The bat-horfes will form and remain on the flanks of their regiments, and the carriages will follow each other, marching in clofe order as directed.

The twelve pieces of cannon attached to each column will file off, betwixt the laft battalion of the firft brigade and the firft battalion of the fecond—The ammunition waggons of thefe pieces will remain in file oppofite the intervals.

The columns ought to be formed in half an hour, from the time the leading battalion or fquadron begins to move.

Upon the fignal being given, the whole army will march at once; and as foon as the columns have quitted the ground on which they were formed, the ammunition waggons and carriages will follow in the order before prefcribed.

### 4.

When the army marches by lines; it will form in column by divifions; twelve fix pounders will draw up in the firft line between the firft and fecond brigades of Britifh infantry, and alfo twelve fix pounders between the firft and fecond brigades of Hanoverian infantry.

## Marechal de Broglie.

When the general beats, the army is always to prepare for moving: the cavalry will faddle, the artillery will harnefs, and the grenadiers, chaffeurs, new guards, quarter mafters, and camp colour men, will affemble one hundred paces in front of the brigades of the firft line.

At the affembly, tents are to be ftruck, cavalry and carriages to load their baggage.

At the troop and to horfe, the whole army will form ready to march, carriages will be placed behind their refpective regiments, and follow in the fame order—The artillery will follow the divifions of the infantry to which it is attached, unlefs directed to take another route.

---

When the troop has beat, and the lines are formed, the infantry will break to the proper flank by platoons, and clofe into the leading regiment of the column. The intervals of platoons will not exceed three paces; and at the fame time, the brigades of the fecond line will join the firft.

When the whole infantry of each column fhall have clofed up, the general officer commanding will order the head to move on.

---

On the *march*, the columns muft always be in readinefs to form; no officer whatever is to ride among the troops,

## Marechal de Broglie.

troops, but to remain and be dreſſed on the flanks of the column. An intelligent one will be advanced about one hundred paces, to reconnoitre the paſſes to right or left of brigades and communications, and to indicate them to the mounted officers. But if obliged to paſs through the ſame defile as the men, they muſt divide themſelves and croſs quickly; half in front, half in rear of the battalion; and when ſo paſſed they will regain the flanks of the columns, and avoid as much as poſſible the neceſſity of entering into them.

---

Excluſive of ſuch pioneers as are appointed to open routes for the columns of the army, fifty per brigade will march at the head of each to repair ſuch bridges and communications, as may have been damaged.

---

The brigadiers ſhall prevent the ſoldiers breaking their files, and attend to their keeping the ſame front they ſet out with. But if ſuch breaking off becomes neceſſary, the defile muſt be paſſed at a redoubled pace, and the diviſions again formed on the other ſide.

---

All operations of doubling and forming up will be taken, and fol-

## Prince Ferdinand.

4. In the ſecond line twelve ſix pounders will draw up between the two brigades of Brunſwick infantry, and twelve ſix pounders between the Heſſian brigades.

The heavy artillery parked in the center of the army will file off, the cavalry will form in column by quarter ranks, and the baggage of each line will draw up on the flanks of their reſpective brigades.

When the ſignal is given, all the columns will begin their march together; and as they quit their ground the ammunition waggons and carriages belonging to the infantry will follow by brigades, according to the order of march in the rear of the column.

The equipage of the cavalry will follow the laſt ſquadrons, and the bat-horſes will march on the flanks of their reſpective battalions and ſquadrons.

No carriage whatever is to march betwixt the diviſions of the columns—The bat-horſes only will be ſuffered to keep on the flanks, but on no account to mix with, advance before, or remain behind their reſpective regiments—The field pieces are to march with their reſpective battalions—The battalions are to keep well cloſed up, and to march by ſubdiviſions.

5.
The commanding officers of battalions, ſquadrons and brigades of artillery, will be reſponſible—That they are formed, tents ſtruck, and baggage loaded in half an hour, from the time that the ſignal for the march was given them; and for this purpoſe, it is neceſſary that they ſhould exerciſe their men to it, while they remain in ſettled camps.

The

## Prince Ferdinand.

The general officers commanding brigades will be responsible—That the columns are formed in half an hour from the time the battalions are drawn up, and the generals who lead columns are to be answerable that they move together exactly at the hour appointed.

6.

The aids de camp, and major of brigades are always to regulate their watches by head quarters, that precision of movement may be attained, and that every brigade may march at the same time.

7.

The guides will always be ordered to the brigades which form the heads of columns, and there present themselves to the commanding officer, of the battalion or squadron that leads.

8.

The commanding officer of each battalion and squadron will pay the greatest attention during the march, and be answerable—That the battalions march always by subdivisions, and the cavalry by quarter ranks—If the *defiles* oblige them to break off, they must double up again as soon as they have passed.

2d. Every officer must remain with his division, and never leave it on any account.

3d. No soldier is to be permitted to quit his rank.

4th. The bat-horses must not be suffered to interrupt the march of the column, but obliged to keep on the flanks.

5th. The distance between divisions is to be exactly observed.

## Marechal de Broglie.

lowed from the front to the rear of the column.

No carriage or bat-horse can ever march in the column, or precede it to camp.

No officer or soldier shall quit his rank without leave.

The cavalry march by troops, and observe the foregoing directions.

Whenever the columns halt, in order to collect and bring up the rear; the infantry will form by quarter battalions, or if near an enemy, by battalions; the cavalry will dismount; centries be posted round each regiment to prevent straggling; and rolls will be called as well as at quitting the old, and entering the new camp; and when to resume their march, the troops must be particularly alert, and in perfect readiness to proceed, otherwise the columns will be again lengthned out, and in worse order than when the halt was made.

As soon as the commanders of columns shall have been informed that they are near an enemy, they will direct and observe as follows.

They will divide the columns they lead each into two columns, or even more if possible, composing them

## Marechal de Broglie.

them of part of the troops of each line.

They will take particular care to obtain and preserve such intervals betwixt the columns, as will enable them all to form in line at one and the same instant.

---

Whenever the troops are near their ground, or by the approach of the enemy they are obliged to form—The battalions and squadrons will close up to each other, preserving a distance not exceeding twelve paces—the officers will dismount, the columns will deploy, and the lines be formed either to right or left.

At the same time that the first line forms, the second line and reserves will also form, keeping or taking a distance of three hundred paces between the two lines—All these movements will be executed as quick as possible, and at a redoubled pace.

The cavalry will conform to the same directions—The brigades of infantry appointed to cover their flanks, will remain in column between the two lines of infantry, dressing with the battalion nearest the cavalry.

---

On the arrival in camp, all staff officers will remain mounted till the tents are pitched, and centries placed; officers of companies will not quit them till those things are done; and the general officers will set an example to the whole.

## Prince Ferdinand.

The generals, or commanding officers of brigades will take care—

1st. That the battalions and squadrons march well closed, and preserve their proper distances.

2d. That the commanding officers of battalions and squadrons punctually obey the orders prescribed—In case of disobedience, the generals are to put in arrest, or correct the disorders of such commanding officers, and report them.

3d. When the army halts, it may be permitted to send for water, but the soldiers must not be suffered to straggle—As many officers or non-commissioned officers as are necessary, must conduct them regularly, and be answerable that no disorder is committed.

4th. No carriages except the cannon are to be permitted to march betwixt the battalions; and in case any of them should stick fast in a deep road, a proper number of men must be ordered immediately to draw them out.

If a carriage breaks, it must be drawn aside, the road cleared, and a proper escort left with it, that the march of the column be not interrupted—The officer under whose care it is left, must get it repaired, and follow the column as soon as possible.

5th. The general officers commanding brigades shall remain with them, and punctually observe the order of march, and the execution of every article prescribed.

---

The generals who lead columns, are to enforce obedience with the utmost severity.

2d. They are to begin carefully their march precisely at the hour appointed; to keep an equal pace, and so to regulate it, that the troops do not exceed three miles in an hour and a quarter.

3d.

## Prince Ferdinand.

3d. The guides serve only to point out the routes for the columns—They must be provided with pioneers to make the necessary openings by bridges, and to repair roads. But the generals must not trust to these precautions, they must gain the most exact knowledge of the routes they are to march, and themselves devise the most proper means to avoid all difficulties that may embarrass the march.

On no occasion can a general officer more display his skill and experience, than in the good conduct of the body which he leads; so that the commander in chief may depend on the exact march of the several columns, and make his calculations accordingly.

Though troops do not always march in presence of an enemy, it is of infinite consequence, that they should always consider themselves as in that situation—Equal and well ordered marches, contribute not only to the preservation of the army, but accustom the troops at all times to be prepared to attack, or repulse the enemy.

### 9.

The quarter master general of each nation, will go forward with the quarter masters and camp colour men—The quarter master general of the army, or such other person as is ordered to reconnoitre the new camp, will direct in general the ground that the brigades of each nation are to occupy; after which the quarter masters general will mark out their respective encampments.

Each nation is to take care to park the brigade of six pounders that is attached to it.

The heavy artillery must detach an officer with their camp colour men to take up the ground of their encampment.

### 10.

Whenever the baggage is ordered to be sent away, all carriages whatever without distinction, are comprehended in this order, except such as may be particularly specified.

The equipage of each battalion or squadron, will assemble in the rear of the camp, exactly at the hour appointed. Those of each

## Prince Ferdinand.

each brigade, on the signal being given, are to file off to the rendezvous of their nation, from whence they will be conducted to the place appointed for them.

The guides ordered to show the way, will come to the commanding officer of the last battalion or squadron of the last brigade of each nation; where the officer who is to conduct the brigade, will come to receive him.

It has been already ordered, that each brigade shall appoint an officer to conduct the baggage; the eldest of whom shall take the charge of the whole of that nation, and be answerable for the departure of it at the proper time, and for the execution of the necessary orders during the march.

### 11.

The chasseurs and light troops, always form the advanced and rear guards, or march on the flanks of the army.

The general officer who commands them, will likewise have under his orders, those battalions or squadrons, that may at any time be ordered to re-inforce the advanced or rear guards. This general officer will always be particularly named.

### 12.

Reserves and detached corps will be regulated according to circumstances, by the above orders.

---

Such were the FIXED PRINCIPLES that directed the movements of very large armies with the greatest honour to their commanders, and in countries offering every variety of surface.

The farther measures necessary for the security of the march; the strength of advanced, detached, and flank corps, and the combined operations of the whole through vast and extended regions, arose from a perfect knowledge of the scene of war, and the skilful arrangements of the general, varied and adapted to circumstances.

Although every necessary appendage to an army may be liberally furnished, yet the foresight and dispositions required to provide for, and to move such bodies to advantage, demand an exertion of the most eminent talents. From these difficulties chiefly, the fortunate general is curbed in his views; and his progress being much regulated by the subsistence of his army, conquest is thereby impeded, the vanquished have time again to make head, and Providence seems in this manner to set bounds to the ravages of mankind.

In order still farther to exemplify the great *movements* and *positions* in the field, we shall here add an imperfect OUTLINE of the CAMPAIGNS of the BRITISH ARMY in GERMANY; but including a more particular detail of the operations that took place prior to the *battle* of *Fellinghusen* in 1761; as a period most critical and interesting, and which produced all the varieties of movement that could be executed by a great army. The very slight manner in which these general events are meant to be touched, will sufficiently apologize for the execution, and must lead us to regret, that no person who may be in possession of sufficient local knowledge and materials, has as yet furnished any detailed or instructive history of a war, fertile in great actions, conducted with superior science, and finished with so much glory.

## Outline of the Campaigns of the British Army, in Germany.

### 1756.

In consequence of the extensive war which had now raged for some time, the FRENCH determined to invade the king's dominions; and at the end of this year, marched one hundred thousand men, under the command of the Marechal D'ETREES towards

*March of the French army to the Rhine.*

wards the Rhine, that they might be enabled to enter lower Germany, either through the countries of *Hesse* or of *Munster*, which are the two great passages that lead into it; and which are separated from each other by the intermediate mountains of the Wetteraw, of the county of Marck, and of the duchies of Berg and Westphalia.

1757.

<small>Passage of the Rhine.</small>

Having determined to advance from the Lower Rhine, the French army in the month of March crossed that river, and took possession of Wesel; which, though very strongly fortified, was evacuated by the Prussian garrison who retired upon Lipstadt, and abandoned this place also, when in the course of the next month the enemy pushed forward a strong corps upon Munster and Ham.

<small>Assembly of the king's army.</small>

The KING's German army, under the command of the Duke of CUMBERLAND, and consisting of between forty and fifty thousand men, Hanoverians, Hessians, Brunswickers and Prussians, began to assemble near Hameln; and in the month of May advanced to the neighbourhood of Paderborn and Bielefeldt; thereby occupying the passages of a ridge of woody hills, which traverse the country, and endeavouring to cover Hesse and Hanover.

<small>Progress of the French.</small>

But Marchal d'Etrees, who on the last day of May had assembled the whole French army at Munster, marched upon Bielefeldt in the begining of June, and the Duke giving way to superiority of numbers, crossed the Weser on the 16th and fell back upon Hameln. Before the middle of July, Embden, Munden, Cassel, Gottingen, and all their dependencies were in possession of the enemy, who on the 10th of that month passed the Weser, and advanced upon Hameln, where in the indifferent position of Hastenbeck the Duke was determined to make a stand.

The

The loss of that battle, on the 26th of July, occasioned his retreat upon Minden, Nienburg, Verden, and Stade successively——Hanover, Brunswick, and Wolfenbuttle surrendered before the middle of August; and on the 8th of September the convention of Closter Seven was signed, which confined the king's army to a small district upon the Lower Elbe, and put the enemy in possession of all Westphalia, and part of Saxony.

<small>BATTLE OF HASTENBECK.</small>

About this time D'Etrees was recalled, and Marechal de Richelieu was sent to command the French army; part of it was put into quarters of cantonment in the conquered countries, and the rest marched towards Halberstadt to re-inforce Prince de Soubise, who with a considerable corps, joined to the army of the empire, was destined for the conquest of Saxony.

<small>Disposition of the French army.</small>

Things remained for some time in this situation, when the king determined that his troops should again take the field, but under a new denomination.—To this he was induced by a variety of infractions of the convention; by the general indignation expressed at a neutrality which tied up the hands of so great an army; by the oppressions, and exactions exercised in the country; by the perishing and helpless condition of the enemy, who were scattered in their quarters, reduced by sickness, without discipline or magazines, and destitute of many of the necessaries of war; by their defeat at Rosbach, and by the alteration which the subsequent victory at Lissa obtained over the Austrians, made in the fortunes of the KING of PRUSSIA.

<small>Re-assembly of the King's army.</small>

They now became the army of Great Britain, were taken into her pay, and PRINCE FERDINAND of BRUNSWICK assumed the command.—About the middle of November he began to act; Harburg was invested, and the quarters of the troops were considerably enlarged; the enemy was obliged to evacuate Luneburg, and Halberstadt, and to assemble in a very shattered condition in the neighbourhood of Zell, there to dispute the passage of the Aller.— About the end of December Harburg surrendered; but the rigour of the season, and the arrangements necessary for pursuing the war, suspended for a time the further operations on both sides.

<small>Movements of the armies.</small>

1758.

1758.

**Movements of Prince Ferdinand, and of the Prussians.**

About the middle of February Prince FERDINAND, joined by fifteen Prussian squadrons, again took the field; Prince Henry of Prussia with fifteen thousand men began to act towards Halberstadt; he advanced to Hildesheim, and then struck off into Saxony.

The French army now commanded on Richelieu's recall, by the Comte de Clermont, apprehensive of being surrounded by these combined movements, retired in the greatest confusion; leaving baggage, artillery, hospitals, &c. a prey to the victors—Bremen was evacuated, and the HEREDITARY PRINCE of Brunswick took Hoya by assault, which secured a passage over the Weser: Minden surrendered on the 14th of March, and in the course of this month the French evacuated all the conquests of the preceding campaign, and retired upon Wesel and Hanau—Much harrassed, and weakned, they were put into quarters of refreshment, betwixt the Rhine and the Maese, while Soubise's army and the corps which had abandoned Hesse, retired behind the Mayn.

**Retreat of the French army.**

**Progress of the British army.**

Prince Ferdinand who had thus pressed the enemy, advanced on the 7th to Dorsten, and then placed his army in cantonments round Munster—In this manner with infinite loss, and for want of discipline, and having taken proper precautions for subsistence, and security in their quarters, was the French army obliged to fly before one inferior in numbers, and to abandon in the most disgraceful manner all their conquests.

**BATTLE OF CREVELDT.**

To improve the advantages already gained, and to profit from the debilitated state of the enemy; Prince Ferdinand after having given some respite to his troops to recover from the fatigues of a winter campaign, again took the field in the end of May.—On the 2d and 3d of June he passed the Rhine below Wesel in the face of a great army, and on the 25th by superior skill and conduct gained a compleat victory at CREVELDT.

The

The French having retired towards Nuys, and Cologne—Detachments were pushed on to Ruremonde, contributions were raised in the Queen's, and bishop of Liege's country, Duffeldorff which secured a passage over the Rhine, was bombarded and surrendered on the 8th of July, and for some time the enemy was not in a condition to make head against the king's army—In this situation, and on the appearance of the seat of war being transferred so near to the Frontiers of France, it was resolved to send a corps of British troops to join Prince Ferdinand's army, one English battalion being already in garrison at Embden, which had been retaken in the month of March, by the assistance of a small squadron of ships. *Consequences of the battle of Creveldt.*

But notwithstanding the brilliancy of the victory at Creveldt, it became very apparent, that it would be impossible to carry on the war beyond the Rhine.—Wesel was in possession of the enemy, and too strong to be attacked; the French army was daily receiving reinforcements; M. de Contades had taken the command of it, and was preparing to act with that vigour which his superior numbers required of him. *Situation of the French.*

Soubise also who had assembled a considerable army upon the Mayn, was now marching through Hesse upon Prince Isemburg, who had been left to cover the entrance into that country, with six or seven thousand men, and who had advanced in the beginning of June to Marburg. On the approach of the enemy he fell back upon Cassel, and near it occupied the strong post of SANDERSHAUSEN, where on the 23d of July he was attacked by the Duc de Broglie, sustained by Soubise's army, and obliged to retire, with loss, upon Hameln, leaving Gottingen and Eimbeck to the enemy. *Progress of Soubise in Hesse, and Affair of SANDERSHAUSEN.*

This advantage, which gave the French possession of all Hesse, and exposed the Bishopricks, the country upon the Weser, and the communication of the army to their inroads, determined Prince Ferdinand to repass the Rhine as soon as possible. But the rains, which had swelled the river to a great degree, damaged the bridge, and left the army in a critical situation, prevented the immediate execution of this plan. *Determine Prince Ferdinand to repass the Rhine.*

Lieut. general de Chevert, deſtined for, but hindered by the ſame reaſons from undertaking the ſiege of Duſſeldorff, reſolved to take advantage of this circumſtance, and to attack lieutenant general Imhoff, who was poſted at MEER, near the confines of Holland, with about four thouſand men to cover the magazines and bridge over the Rhine, and who then had a very imperfect communication with the king's army. He marched from Weſel with a much ſuperior force; but on the 5th of Auguſt was met and attacked by Imhoff in a critical moment, before he had compleated the formation of his troops; they were put to the rout with the loſs of a conſiderable number of men and of ſeveral pieces of cannon.——This fortunate action enabled Imhoff to ſecure and repair the bridge, to keep up his communication with the army, and to enſure the junction of the BRITISH corps, conſiſting of ſix battalions and 14 ſquadrons, which under the command of the Duke of MARLBOROUGH, had landed at Embden the end of July, and was at this time on the borders of Frieſland.

*Affair at MEER.*

A defeat in this critical ſituation of the army, conſidering the ſtrength of Contades, and the progreſs of Soubiſe, might certainly have been attended with the worſt conſequences.—Prince Ferdinand would moſt probably have been obliged to paſs the Rhine at Duſſeldorff; and with loſs he muſt have found very great difficulty in regaining the country of Munſter, or in joining the Britiſh corps, which might have been forced back upon the Ems, or Weſer. But the enemy diſpirited by their ill ſucceſs, gave no diſturbance to Prince Ferdinand, when he paſſed the Rhine on the 9th and 10th near Rees. Duſſeldorff was at the ſame time evacuated, and the garriſon retired by the Roer into the county of Marck, without any loſs. The whole army joined in the camp at Coesfeldt, on the 25th of Auguſt, where the Britiſh corps of 8000 men had been ſince the 14th.

*Prince Ferdinand repaſſes the Rhine.*

Contades, now made a Marechal of France, croſſed the Rhine on the 20th near Weſel, with his whole force; but though much ſuperior in numbers, he was not able to advance into the country, or to undertake any operation of importance. He was obliged to draw his

ſubſiſtence

subsistence from the Rhine, and from Wesel, where no magazines could have been previously established; nor could he move on, without rendering his communications too precarious, while Prince Ferdinand remained in the position of Coesfeldt and Dulmen; where having the countries of Munster, Friesland, and the navigable river of the Ems behind him, he was abundantly supplied with whatever was essential.—The harrassing service of the early part of the year might contribute to that inaction in which the armies remained during the months of August and September, for there was no considerable change in their position: the Lippe divided them; Prince Ferdinand kept his camp at Coesfeldt and Dulmen, and Marechal Contades continued near Recklinghusen. On the death of the duke of Marlborough at Warendorff, lord G. Sackville had succeeded to the command of the British troops.

*Operations of Contades.*

After the affair of Sandershausen, Soubise who had only sent a small corps towards Gottingen, marched with the greatest part of his force to Warburg, and made detachments towards Lipstadt and Paderborn. On the 8th of September he crossed the Weser with his troops, to make an irruption into Hanover, but was obliged on the 26th to return to Cassel to oppose lieutenant general Oberg, who, detached from the army with a considerable force, had encamped the end of August under Lipstadt, and having been joined by Prince Isembourg, now advanced upon Cassel in the absence of Soubise.

*Operations of Soubise and Oberg.*

To favour these movements in Hesse, Contades, who about the time of the junction of the British, had been reinforced by ten thousand Saxons, taken into the pay of France, moved about the 23d to Ham, drove Kilmansegge into Munster, and detached thirty-five battalions and thirty squadrons to reinforce Soubise, while Prince Ferdinand still remained at Dulmen. Oberg who had encamped for some days before Cassel, crossed the Weser on the 5th of October, and took post at Landwerhagen, where he remained till the 9th, when he fell back upon LUTERNBERG, and was there attacked on the 10th by Soubise reinforced from Contades's army. With the loss of above three thousand men he was obliged to retreat upon Munden.

*Battle of LUTERNBERG.*

To

( 214 )

**Movements of Prince Ferdinand and Contades.**

To assist Oberg, and to profit of the opportunity of attacking Contades, while weakened by the great detachment sent into Hesse, Prince Ferdinand marched from Dulmen on the 7th of October, and after remaining some days at Munster, passed the Lippe near Lipstadt, drove in the enemy's advanced corps at Soest, and on the 20th moved forward to Werle, where he found the enemy on the heights of Wambeln, behind the Saltzbach (the position occupied by the Hereditary Prince's corps, at the battle of Fellinghusen in 1761) too strongly posted to be attacked. He was therefore obliged to give up this intention and to return to Lipstadt, where Oberg came with the greatest part of his corps, and from whence Prince Ferdinand proceeded to his camp at Munster. Contades was joined on the 23d by his detachment from Soubise, and remained in the neighbourhood of Ham.

**Close of the campaign.**

This was the last operation of the campaign, and the French, notwithstanding their success and superiority of numbers, not having acquired any post of sufficient strength, which they could provide and maintain during the winter, saw themselves again obliged to relinquish their conquests, and to approach their own frontiers.

**Winter quarters.**

About the middle of November Soubise evacuated Hesse, and went into winter quarters behind the Mayn. Contades also marched upon Wesel, crossed the Rhine, and placed his army between that river and the Maese. The allied army broke up the camp at Munster, November 15th, and took their winter quarters in the countries of Hesse, Munster, Osnaburgh, Paderborn, and Hildisheim—Marburg and Ziegenhayn were the advanced posts towards the army of Soubise; and Ham, Coesfeldt and Dulmen, towards that of Contades.

1759.

On the 2d of January the French took possession of *Francfort* on the Mayn; and in the next month the army of the empire quartered in Thuringia, made an irruption into Hesse, as far as Hirschfeldt;

the

the French light troops acting in conjunction, advanced to Marburg, but on the approach of major general Urff, who was detached for that purpose, they were obliged to retire. About the middle of March the army of the empire again extended themselves into Hesse, towards Fulda and Hirschfeldt, while strong French detachments pushed on towards the Lahn, and the Ohm. The principal object of these movements was to impede the recruiting of the Hessian troops, and prevent the collection of forage for the magazines. *{Movements of the enemy in Hesse.}*

Prince FERDINAND, in order to check this progress, had assembled a considerable part of his army in the neighbourhood of Cassel, and marched upon Hirschfeldt, while the Prince of Holstein, with a corps, advanced upon Marburg. The enemy was obliged to retire, Prince Ferdinand was on the 27th master of Fulda; and the army of the empire hemmed in on the other side by the Prussians in Thuringia, were obliged, with considerable loss, to make their retreat upon Nuremberg. *{Movements of Prince Ferdinand.}*

Prince Ferdinand, who had thus got rid of these invaders, and who had now collected a considerable corps, consisting of all the Hessians and Brunswickers; seven English, and ten Prussian squadrons; seven battalions and seven squadrons of Hanoverians, besides light troops: after remaining some time at Fulda, marched upon M. de Broglie, who had assembled the French army of the Upper Rhine in the strong position of BERGEN, near Francfort. Prince Ferdinand, to whom a victory would have produced the greatest consequences, by giving him the possession of Francfort, of the Mayn, and by the probable retreat of the French from that part of Germany; and who, if unsuccessful, would only the sooner quit a tract of country where he had no establishment, and which he must necessarily relinquish if no action took place, attacked Broglie on the 13th of April, but was repulsed with considerable loss, and retired upon Windecken, without being pursued by the enemy. Broglie, immediately after the battle, was joined by a large body from the Lower Rhine, under M. de St. Germain; the army of the empire began to *{Battle of BERGEN.}*

act

act upon his right towards Fulda, and Prince Ferdinand continued his retreat gradually into Hesse, arriving at Ziegenhayn the 23d.

While these operations were going on in Hesse, Imhoff had commanded a corps assembled near Lipstadt, and the quarters of the rest of the army, in the country of Munster, had been contracted to oppose any attempts which the French troops on the Lower Rhine might have made in Prince Ferdinand's absence.—In the end of April, Imhoff with his corps was sent towards Cassel, and the army which had acted in Hesse, went into quarters of refreshment, in the countrys of Paderborn and Marck; while the French did the same about Francfort, Giessen, and in the Veteravia.

*Cantonment of the army.*

The enemy, from the experience of last year, having seen the advantage which the possession of the course of the Lippe, Lipstadt, and of Munster, where large magazines were established, gave to Prince Ferdinand, in defending that entrance through the flat country into Westphalia; now determined to make their principal effort by the other through Hesse, which was more woody and mountainous, and abounds in strong positions.—Their army of the Lower Rhine encamped in several corps the 6th of May. Detachments were pushed into the Veteravia, to cover the march of a great proportion of it, which in several divisions crossed the country between the Lahn and the Rhine, and on the 31st was assembled, together with the army of the upper Rhine, in the neighbourhood of Giessen—Marechal de CONTADES took the command, while D'Armentieres remained near *Wesel* with about twenty thousand men.

*Assembly of the French army.*

On the 10th of May the Hereditary Prince, with a large corps, had advanced towards Fulda, but was recalled on the march of the French army towards the Lahn. Imhoff, with ten or twelve thousand men, encamped near Fritzlar, and Prince Ferdinand, with the body of his army, remained cantoned on the Lippe, and in the county of Munster, as in a central position, till it could be determined from whence the great effort of the enemy was to be made; whether from the Mayn, or from the Lower Rhine. The assembly of the French army on the Lahn at last determined the removal of the magazines,

*Position of the British army.*

those

those of Munster to Lipstadt, those of Osnaburg to the Weser, and the march of the greatest part of the army from the country of Munster to the neighbourhood of Ham, where it encamped the begining of June in different corps, and on the 8th the whole assembled in the camp of Werle.

The French army preceded by its advanced corps was put in motion the 3d of June, Marburg was occupied the 5th, Ziegenhayn the 6th, and while Broglie pushed on to Cassel which he entered the 11th, Contades with the army arrived on the 10th at Corbach.—On the 14th, having gained the passes of Stadberg he crossed the Dimel at that place, while Broglie on the same day was at Warburg, and on the 16th, after marching over the Paderborn hills, he arrived at Leichtenau within two leagues of Contades—D'Armentieres who crossed the Rhine on the 15th, with nineteen battalions and twenty squadrons, advanced to Schermbeck four leagues from Wesel where he remained for the rest of the month, and was watched by lieutenant general Kilmansegge who encamped at Dulmen with a small corps, to cover Munster. *Progress of the French armies.*

While Contades was making these rapid advances; Prince Ferdinand who on this occasion does not seem to have been sufficiently apprized of the motions of the enemy, or may not then have fully reflected on the great importance of the positions of the Dimel, did not arrive at Buren five leagues from Stadberg with the head of his army till the 14th, too late to oppose the passage of the defiles which Contades had now gained.—At Buren he was joined by the greatest part of Imhoff's corps which had retired before Broglie upon Cassel and Warburg, after sending off a small detachment towards Gottingen to cover that country from the incursions of the enemy's light troops: the army now consisted of fifty battalions, fifty-nine squadrons and light troops. *Movements of Prince Ferdinand.*

Contades being in possession of all Hesse and having assembled his army of 110 battalions and 106 squadrons, in the country of Paderborn, began to extend on his right, in order to encompass Prince Ferdinand's left flank who was posted behind the little river Alme, and had a small corps on the other side of the Lippe. *Movements of Contades.*

In the situation of Buren though Prince Ferdinand might have been able to keep up his communication with Lipstadt, that with Munster would have been precarious; as d'Armentieres if enterprising, was able to drive Kilmansegge back upon it.—The nature of the country prevented his effectually disturbing the communications of the enemy, and in case of a battle, he would have had the whole superior French force to combat, fresh, inspirited with their success, and not yet weakned by detachments or garrisons.—At any rate he left Hanover, and the whole country on the Weser at their discretion, where his essential posts were such as could make no considerable defence.—Since the beginning of the war the fortifications of Munster had been considerably strengthned, and those of Lipstadt particularly so, it was now a respectable place.—Throwing therefore a sufficient garrison into Lipstadt he marched the 18th at night, crossed the Lippe, and encamped on the 21st at Ritberg.

*Necessity of Prince Ferdinand's retiring.*

Contades now advanced to Paderborn where the necessity of bringing up supplies for his army kept him some time, but on the 29th he moved on to Lipspring.—This determined Prince Ferdinand to quit his position of Ritberg, and to march by Guterslohe and Marienveldt upon Dissen, where he encamped the 3d of July.—Contades moved the 2d to Stuckenbrock, and on the 4th to Bilefeldt: Broglie's reserve was near Hervorden, which was occupied by his light troops; and Chevreuse's reserve of the left was placed at Delbrugge.—On the 7th Broglie's reserve was advanced to Engern, and on the 8th Contades marched to Hervorden, and Chevreuse to Ritberg, leaving Lipstadt blockaded.

*Progress of Contades.*

These movements of Contades which still tended to gain Prince Ferdinand's left flank, obliged him on the 8th to march to Osnabruck, and to recall a strong detachment which he had sent to Melle.

Armentieres who had advanced and driven Wangenheim back upon Munster, invested that place on the 9th—Broglie on the same day carried Minden by surprize and assault, his reserve encamped under

*Minden taken.*

under it on the 11th, and on the 14th croffed the Wefer, being replaced by another corps.

Contades remained near Hervorden to obferve Prince Ferdinand, who having failed in his intention of drawing the enemy to a diftance from the Wefer into the bifhopricks, and now feeing himfelf in danger of being cut off from that river fince the lofs of Minden, was obliged to quit his pofition near Ofnabruck on the 11th, and march by Bomte and Raden to Stoltzenau, where he arrived the 14th, in order to cover his magazines and hofpitals at Nienborg, and Bremen.

*Retrograde movements of Prince Ferdinand.*

On the 14th Contades quitted Hervorden, and on the 15th encamped at MINDEN—St. Germain who had been for fome time detached towards Caffel arrived on the 16th near Hameln, and Broglie with his referve repaffed the Wefer in confequence of Prince Ferdinand's movement from Stoltzenau to Peterfhagen on the 16th, but he again paffed it on the 17th.

*Contades encamps at Minden.*

On the 17th Prince Ferdinand advanced into the plain of Minden with his army, and after examining the French pofition returned to his camp of Peterfhagen; but Wangenheim with his corps remained and intrenched himfelf behind Thonhaufen, and the Piquets and pofts of the army were extended on the right quite to the Marfh of Minden—In this fituation the armies remained for fome time—Contades was employed in fecuring his communications which were now very extended, and in bringing up his fupplies from Paderborn, at the fame time that he waited the event of the fiege of Munfter which D'Armentieres took on the 25th after fix days open trenches—Prince Ferdinand was employed in reconnoitring and making preparations for giving battle to the enemy, which he faw he could do to advantage, whenever they quitted their ftrong fituation.

*Prince Ferdinand encamps at Peterfhagen.*

This camp of Minden, which might be advantageous for a fmaller corps, and was indeed inattackable, was a very improper one for fo great an army acting on the offenfive, to be cooped up in. With a chain of woody mountains immediately in the rear

*Nature of the camp at Minden.*

and

and on the left, a river upon the right, and a morafs extending along the front; there was no means of advancing from this pofition, but by marching along the narrow fpace between the town and morafs, and afterwards forming in line when arrived in the plain of Minden. That this fituation muft foon occur, Prince Ferdinand forefaw—to attack the enemy while in the act of extending themfelves, or before their columns fhould have completely croffed the narrow ground, was his fixed and early intention. That he afterwards was not timeoufly enough apprized of their motions, was the fault of his advanced pofts, not of his difpofitions. Every arrangement made for many days before the battle, had this defired object in view; all neceffary openings were made, and routes traced for the expeditious march of the army; every thing was prepared which human forefight could fuggeft; and every individual kept ready to feize the expected opportunity.

When Prince Ferdinand had quitted Ofnabruck, the French put a fmall garrifon into it, and one alfo into Hervorden to cover their communication; a confiderable corps encamped at Coveldt, on the river Werra, where they had eftablifhed their bakeries; they alfo took poffeffion of Lubbeké, at the head of the morafs of Minden, and a pafs through the hills. In order to difturb this communication, Lieut. Gen. Dreves with fome light troops, and the greateft part of the garrifon of Bremen, marched upon Vecht, a fmall poft which the enemy held invefted, and advanced to Ofnabruck which he took by affault, the fame day that the Hereditary Prince detached from the army with about fix thoufand men, forced the poft of Lubbeké. He joined Dreves on the 29th, on the 30th was near Hervorden, and on the 31ft placed himfelf at Quernum near Coveldt, with eleven battalions, eleven fquadrons, and three of huffars.

*Situation of detached corps.*

To protect thefe movements, made in a fituation fo critical, Prince Ferdinand on the 29th, marched his army by its right to Hillé, placed that flank near the morafs, and with its left approached Wangenheim, who remained behind Thonhaufen where he had thrown up confiderable works. On the 30th, Lieut. Gen. Gilfæ,

*Camp at Hillé and Thonhaufen*

Gilfæ, with three battalions and a detachment of cavalry, took poſt at Lubbeké—The piquets of the infantry and cavalry, were puſhed as forward as poſſible on the plain of Minden to give the earlieſt intelligence.—Orders where given that the generals ſhould reconnoitre the ground, and make themſelves maſters of the routes of the columns—The army was directed to be in readineſs to advance on the firſt notice of the enemy's being in motion, and to take up a given poſition on the plain, the right at the village of Hahlen, the left extending towards Wangenheim's corps, whoſe left touched the Weſer. In this ſituation, Prince Ferdinand was ready either to give battle to Contades if he advanced, or to aſſiſt the Hereditary Prince in his intended attack upon the Duc de Briſac at Coveldt. Both theſe fortunate events took place on the ſame day, the 1ſt of Auguſt.

Contades from the abſence of the Hereditary Prince, and from the extent of ground which the Britiſh army now occupied in its two camps of Hillé and Thonhauſen, thought he had an opportunity of attacking the latter to advantage. In the night of the 31ſt, Broglie repaſſed the Weſer; and the whole French army, conſiſting of ninety battalions and ninety-four ſquadrons, marched at midnight in nine columns, and was formed in order of battle about five next morning, on the plain of MINDEN; their center and left being nearly advanced to the poſition marked out, and intended for that of the king's army of forty-three battalions and fifty-eight ſquadrons. *March of the French army.*

Although the troops had been ordered to be in readineſs to march at one in the morning, yet Prince Ferdinand and Wangenheim, not properly apprized of the enemy's movements, were not in motion before five o'clock—The right of the army however, advanced towards its ground; about half paſt ſeven the battle began; about nine the French gave way, and before ten they retired in the greateſt diſorder to their old camp, behind the moraſs; having loſt many cannon, ſtandards, colours, and near ſix thouſand men. The loſs on the ſide of the allies, did not amount to one thouſand five hundred men. The Britiſh infantry on this occaſion, *BATTLE OF MINDEN.*

high'y

highly diftinguifhed themfelves in withftanding and repelling the repeated attacks of the enemy's cavalry, and the gallantry of the troops in general, feconding the wife difpofitions of their commander, obtained a moft fignal and brilliant victory, which would have been compleat indeed, had the cavalry as ordered, been led on with that rapidity, which the critical minute of fuccefs required.

But what rendered this day the more decifive in its confequences, was the defeat, which at almoft the fame hour the Hereditary Prince gave the Duc de Briffac, near COVELDT; and the deftruction of the two bridges over the Werra. This effectually cut off the retreat of the French army upon Hervorden, which they had begun to make, 'till informed of the difafter by the fhattered remains of Briffac's corps, who were driven in upon them, and obliged them that fame night to crofs the Wefer, and retire in the greateft diforder towards their magazines in Heffe.

*Action at COVELDT.*

On the 2d Minden furrendered, and the Hereditary Prince with a confiderable corps, followed the enemy; made feveral attacks upon them, took many prifoners, much baggage, and purfued them by Oldendorp, Haftenbeck, Eimbeck, Gottingen, quite to the defiles of Munden which they paffed on the 9th. All their detachments quitted the bifhopricks; D'Armentieres raifed the blockade of Lipftadt, which he was preparing to befiege; fent back his artillery to Wefel, and himfelf marched, after putting a ftrong garrifon into Munfter, with the reft of his corps towards Warburg. Contades' army was affembled on the 11th near Caffel; and D'Armentieres, who joined him in that neighbourhood, was from thence pofted near Wolfshagen.

*Retreat of the French.*

Prince Ferdinand, after refting his army fome days at Minden, marched on the 4th to Coveldt, 5th to Hervorden, 6th to Bielefeldt, 8th to Stuckenbrock, and on the 9th to Paderborn. To oblige the enemy to quit Caffel, where they feemed determined to make a ftand; he marched on the 11th to Dalem; on the 13th he croffed the defiles of Stadberg; and on the 17th encamped at Arolfen.

*Movements of the armies.*

This situation alarming the French for their communication with Franckfort, they marched on the 18th to Fritzlar; and on the 22d were encamped behind the Ohm, with D'Armentieres at Marburg, and Broglie's reserve towards Ziegenhayn, from whence he afterwards fell back opposite to Wetter.

Prince Ferdinand advanced on the 19th to Corbach, 21st to Frankenberg, and on the 28th encamped at Oberwetter, after dislodging the enemy, whose main army was encamped behind Amœneburg. Marechal D'Étrees arrived about this time to assist Contades; and D'Armentieres returned to the Lower Rhine to assemble the troops left in that country, and others who where on their march from France. The Marquis of GRANBY now commanded the British corps.

<small>Progress of Prince Ferdinand.</small>

Such were the immediate consequences of the battle of Minden, aided by intrigue and disunion in the French army, the want of supplies, and of a strong well provided post on which to retire— The enemy were now considerably diminished in their numbers, their hospitals and magazines were in the hands of the victors, discontents prevailed among them, while the most perfect confidence in the abilities of their commander reigned through the British army. Conquests so rapidly made, were as rapidly relinquished; and Munster, Marburg and Ziegenhayn, were the only acquisitions of this year that remained in their hands, and of which it was necessary that Prince Ferdinand should dispossess them, in order to insure winter quarters to his army.

<small>Consequences of the battle of Minden.</small>

Still persevering in his operations on the enemy's left, at the same time that the light troops acting towards the Fulda, incommoded their communication on the right; Prince Ferdinand by his movements on the Upper Lahn, obliged Contades to quit his position behind the Ohm; on the 4th of September, he fell back to Muntzlar, and on the 7th encamped near Giessen; the Saxon troops there left him to return towards their own country, where the advantages lately gained by the imperialists, gave them hopes of again repossessing it.

<small>Retreat of the French.</small>

Prince

**Camp of Crofdorff.**

Prince Ferdinand now invested and took Marburg, and on the 18th encamped at Crofdorff, opposite to the enemy at Klein Linnes. The Lahn separated the two armies, who both began to send considerable detachments towards the Lower Rhine, to support their respective operations on that side.

Imhoff, with a considerable corps had besieged Munster since the 24th of August, but was obliged to retire on the approach of D'Armentieres September 5th, who re-victualled the place, and marched back upon Wefel; but being joined by a considerable corps of cavalry from France, he again advanced, on the 24th encamped at Luynen, pushed his light troops to Soest, threw further succours into Munster, and in the beginning of October fell back to Dorsten, where he waited for the troops which Contades was sending to him.

**Siege of Munster.**

Imhoff much strengthned by detachments from the army, blockaded Munster in the end of October; and having received further reinforcements and his heavy artillery, opened trenches on the 9th of November. D'Armentieres advanced to its relief on the 19th; but Imhoff, who covered the siege carried on by Count la Lippe, having destroyed the roads, secured the avenues, and thrown every impediment in his way, obliged him to retire again upon Dorsten; and on the 21st the garrison surrendered.

**Situation of the two armies on the Lahn.**

The two armies still remained in the same position on the Lahn, much occupied and finding great difficulties in procuring forage and subsistence. The cavalry was sent into cantonments, contiguous to their positions; but the infantry remained encamped, hutted, and the whole in a constant state of alertness. The Saxons together with the army of the empire, were placed in the cities of the Mayn from Hanau to beyond Wurtzburg. About this time also, ten thousand Wirtembergers taken into the pay of France, were ordered to march and make a diversion towards Hesse; they arrived at FULDA on the 21st of November, where they cantoned, occupied Hirschfeldt and Vacha with their light troops, and their left was covered by the French posts of Lauterbach,

terbach and Herbstein. In this situation, they remained till the end of the month, raising contributions, and cutting off supplies of all kinds, which the countries on the Fulda had hitherto afforded to the allied army.

As it was necessary to dislodge such troublesome neighbours, the Hereditary Prince, with a considerable corps marched from Marburg on the 28th; on the 29th he forced the French light troops from Lauterbach, and continuing his march, on the 30th in the morning, he surprized the Wirtembergers dispersed in their cantonments, drove them from Fulda with great loss, obliged them to quit the country, and to fall back upon the Mayn in the utmost disorder. *Defeat of the Wirtembergers at* FULDA.

The Hereditary Prince returning from Fulda, was on the 3d and 4th of December at the sources of the Ohm, on the left flank of the French army. This position, and the lateness of the season, induced BROGLIE who for a month past had commanded, to march back upon Friedberg, and canton his troops in that neighbourhood, leaving St. Germain with an advanced corps at Butzbach, and a garrison of two thousand men in Gießen. *Cantonment of the French.*

Prince Ferdinand, who still kept his camp of Crofdorff, now sent several detachments across the Lahn, and formed the blockade of Gießen, but could not undertake the siege, being obliged in consequence of the unfortunate affair of *Maxen* to detach a very large corps under the Hereditary Prince; he marched about the 8th, and surmounting every obstacle the army of the empire could throw in his way, crossed Thuringia, and joined the KING of PRUSSIA at Freyberg, near Dresden, on the 28th. *March of the Hereditary Prince to Saxony.*

On the march of the Hereditary Prince, *Broglie* now made a Marechal of France, advanced with his army, and putting the Wirtembergers in motion on his right, and a detachment from D'Armentieres on his left upon the Upper Lahn, he hoped to oblige Prince Ferdinand to quit his camp of Crofdorff—The posts of Klein Linnes and Langhorn were attacked on the 22d, on the 25th the blockade of Gießen was raised, and on the 27th the allied *Close of the campaign.*

allied army aſſembled and lay au Bivouac.—But Broglie finding it impoſſible, from the rigour of the weather, to continue his operations with any proſpect of ſucceſs, fell back to Friedberg on the 28th.

Prince Ferdinand, who notwithſtanding the remarkable ſeverity of the ſeaſon, had ſo long perſevered in his poſition of Crosdorff, poſſibly from the neceſſity of aiding the Pruſſian operations in Saxony; broke up his camp on the 3d of January, cantoned his troops for ſome days round Marburg, and on the 7th proceeded through Heſſe. Both armies now ſeparated, and remaining in poſſeſſion of the ſame country and poſts as laſt year, took up the ſame general winter quarters. The Hereditary Prince returned alſo from Saxony, and cantoned in Heſſe and on the Fulda.

Winter quarters.

## 1760.

The winter paſſed without any very conſiderable event taking place—But the enemy, who were now commanded by the Marechal de Broglie, determined to exert every nerve, and greatly reinforced in numbers, being expected again to make their chief efforts through Heſſe, Prince Ferdinand prepared to prevent their re-entering that country, and to oppoſe them at the paſſage of the river Ohm.— The *poſition* on the Ohm—viz. behind Homburg, Schweinſburg, Kirchain, with a ſtrong corps on the right at Wetter, on the Lahn, together with the poſſeſſion of Marburg, effectually covers the two entries from the Mayn into the Lower Heſſe, can hardly be turned on the flanks, and is difficult to be forced in front—It was intended to be poſſeſſed at the opening of this campaign; was occupied by the Britiſh army in March 1761, to cover the ſiege of Caſſel; and again in October 1762, at the time the peace took place.

Poſition on the Ohm.

Having left general Sporken in the country of Munſter, to face St. Germain, who commanded a conſiderable army of thirty-two battalions and thirty-two ſquadrons, beſides light troops, on the Lower Rhine; Prince Ferdinand early advanced a ſtrong corps upon Fulda, Homburg, and Marburg, and with the body of the army encamped

Camp of Wavern, 20th of May to 22d June.

encamped on the 20th of May near Fritzlar, within three marches of the Ohm; that the forage of the country might be the longer preserved, and the position the more easily maintained. The German troops had been much augmented, seven battalions and eight squadrons had arrived from England, and farther reinforcements were daily expected, so that the establishment of the army this year amounted to ninety thousand men.

But Broglie having with great secresy assembled his superior army of one hundred and eighteen battalions and one hundred and sixteen squadrons, besides light troops, round Gieſſen; advanced, and on the 24th of June forced the passages of the Ohm at Homburg, before Prince Ferdinand, who had marched from Wavern, could arrive there; drove in his advanced corps, commanded by the Hereditary Prince, and Imhoff who never served after this period; obliged him to retire to Ziegenhayn; took Marburg; and then moving rapidly on the 7th of July to his left, was joined by St. Germain on the heights of *Corbach*, on the 10th, where he that day gained a considerable advantage over the Hereditary Prince, supported by the head of the army. The strong position of *Saxenhauſen*, which Prince Ferdinand immediately saw and occupied, after failing in that of *Corbach*, prevented the enemy for some time from entering Lower Heſſe, or the country of Paderborn, which were now their objects, and nearly open to them.

Broglie paſſes the Ohm —Affair of CORBACH, 10th July.

While such was the situation of the armies at Corbach and Saxenhauſen, M. de Broglie, in order to maſk Ziegenhayn, which we still poſſeſſed, and to cover his communication with Marburg and Gieſſen, from whence he drew his supplies, had left five battalions and the corps of Berchini, horse and foot, encamped at *Emsdorff*, their left to the village and the wood; their right extending into the open country.———Prince Ferdinand, informed of the position of this corps, determined to attack it, although so far removed. On the 14th six battalions of the army marched, and next morning early, the Hereditary Prince aſſembled at Zweſten beyond Fritzlar, those six battalions, Luckner's huſſars, a squadron of Freytag's chaſſeurs, and ELLIOT's regiment

Poſition at Corbach and Saxenhufen.

regiment of six hundred men newly arrived from England, and by forced marches brought to the rendezvous. That evening this corps marched to Ziegenhayn, and next morning about eleven arrived in front of Speckwinckel, where they formed under cover of a rising ground about two miles from the enemy, who had no idea of their danger, and were prevented from discovering it by the advanced parties of chasseurs, which were posted in the front for that purpose early in the morning.

*Affair of* EMSDORFF, *16th July.*

About twelve the Hereditary Prince, taking five battalions and a body of chasseurs, marched to the right, made a considerable detour across the woods and hills by the village of Wolffkulen to gain the enemy's left; while Luckner with the cavalry and one battalion, remained in the front of Speckwinckel, to engage their attention.—About two the Hereditary Prince came out of the wood, formed, and attacked the enemy's left, who were completely surprized, being in the act of delivering their bread, and had only time to place two battalions on that flank—On the first firing Luckner advanced on their right where their cavalry was placed; little resistance was made; the whole went off through the woods: baggage, tents, and artillery were all taken, with some prisoners, and the cavalry was pursued and completely dispersed.

*Attack and surrender of the enemy.*

In the mean time the enemy's infantry being, by the position of the Hereditary Prince, cut off from the direct road to Marburg, collected together, retired through the wood upon Langstein, and meant to have proceeded to Kirchain, there to have passed the Ohm. But Eliott's regiment having in the course of pursuing the cavalry, already placed themselves on that route, this infantry was obliged to change its intention, and by marching through Langstein upon Nider Klein endeavoured to gain the heights of Homburg. Near Langstein their rear was attacked by Eliott's, and some prisoners made; the column however proceeded slowly through intricate ground, while the Hereditary Prince, who found his infantry too much fatigued to follow, made a circuit at the head of Eliott's and the horse chasseurs only, to the right by Blasdorff, and came up with the enemy near Nider Klein, who after giving a very destructive fire were attacked

in

in several places, broke, and about five hundred made prisoners.—The remainder of their column still persevering in their endeavour to gain a wood about a quarter of a mile distant, which would have effectually sheltered them, the regiment again prepared to renew the attack, and had advanced within two hundred yards for that purpose, when a parley was beat, and the whole surrendered about five in the evening: Mr. de Glaubitz who commanded, about two thousand men and officers, and six pieces of cannon, were taken.——Our loss in killed and wounded consisted of one hundred and eighty-six men and one hundred and eighty-one horses, of which one hundred and twenty-four men and one hundred and sixty-eight horses belonged to Eliott's, who on this occasion had brought forward about five hundred men.

The whole merit of the event of this action was due to ELIOTT's regiment commanded by Major *Erskine*, which newly raised, disembarked, and proceeding to join the army, was ordered when on the other side of Cassel upon this service. After an almost uninterrupted march of three days and nights, and when the infantry, exhausted with fatigue, were unable to follow; it continued the pursuit, and under the direction of the Hereditary Prince, attacked, broke, astonished the enemy, and obliged that great body of infantry to surrender.

The affair of *Emsdorff* however honourable for those concerned, was not of sufficient magnitude to influence the operations of the two principal armies.—The superiority of Broglie allowed him to detach on both flanks of his adversary, who after maintaining in the most wonderful manner the accidental position of *Saxenhausen* for above a fortnight, was at length obliged to relinquish it, and to fall back upon Cassel.—In full expectation of forcing the British army over the Weser, and into Hanover, he detached Du Muy who now commanded St. Germain's corps of thirty thousand men, across the Dimel to *Warburg*, still more to circumscribe us on the right.

*Prince Ferdinand falls back to Cassel.*

Prince Ferdinand sensible of the danger of being cut off from the bishopricks and confined in a narrow space, saw there was no farther time to be lost in extricating himself; he marched in the evening

evening of the 30th by his left towards Caffel; but as foon as it was dark, his columns changed their direction, moved to the right, croffed the Dimel, and near Liebenau joined the Hereditary Prince and Sporken who had been previoufly fent forward with a large corps, and were deftined to begin the battle and turn the left of the enemy.—The whole marched on in five columns, a foggy morning favoured the approach of the army, and about mid-day of the 31ft, Du Muy, formed on the heights of Warburg, unfupported by Broglie, and not imagining he had fo great a force to deal with, was attacked and defeated: with the lofs of feveral cannon and three thoufand men, he was driven over the Dimel and obliged to fall back upon Broglie, who when too late was moving from Caffel to his affiftance.

<small>Battle of WARBURG, 31ft of July.</small>

This great advantage, and the ftrength of the extenfive pofition which Prince Ferdinand now occupied behind the *Dimel*, enabled him to cover the bifhopricks of Munfter, Paderborn, and Ofnabruck of which he remained in poffeffion, whofe communication with Holland and the fea was fo effential to his fubfiftence, and whofe prefervation he was determined never to lofe fight of. *Hanover* remained protected by the difficulties of the country, the woody mountains of the Hartz, and a fmall corps under Kilmanfegge and Wangenheim which had evacuated Caffel and Gottingen, and was afterwards pofted on that fide of the Wefer at Uflar during the remainder of the fummer.

<small>Pofition of the Dimel.</small>

Marechal de Broglie having feen the danger of too rapid an advance, and that folid eftablifhments alone could fecure him the poffeffion of the country he had acquired; inftead of endeavouring to grafp at greater advantages, and ftill farther to pufh his fuccefs, turned his whole attention and arrangements towards the beginning of autumn, to the fortifying of Caffel and Marburg, as points by which together with Ziegenhayn and Gieffen he meant to hold his prefent poffeffions, and to keep up his communication with the Mayn.

<small>Defigns of Broglie.</small>

Prince Ferdinand who had trufted that the enemy, as in former years, would be obliged in the winter to evacuate Heffe, and again to return to the Mayn and the Rhine; on finding this change of fyftem,
endeavoured

endeavoured by every means in his power to retard its execution and prevent its taking place. Although he could not directly attack Cassel, and impede these operations, he hoped to effect it by a powerful diversion, and towards the end of September he gradually detached a very large part of his army to assemble on the Lower Rhine, which the enemy had left totally unguarded.—The light troops crossed that river, and scoured the whole country beyond it; the Hereditary Prince took the command; *Wesel*, which was garrisoned by one thousand men only, was invested; battering cannon were ordered up from Munster, C. la Lippe had the conduct of the siegé, a bridge of boats was established immediately below the town, Cleves and Rhinberg were occupied by the light troops, and intrenchments were thrown up on the opposite shore to Wesel. The undertaking however exceeded our force, although forty-seven battalions and thirty-two squadrons marched and were destined for the enterprize; the difficulty of transporting artillery and stores prevented our making any great progress before the town, and the extent of the operation proved to be beyond our ability.

*Siege of Wesel.*

Lieut. Gen. de Castries was now assembling an army about Cologne, composed of troops from France, and of large detachments successively sent from M. de Broglie with great expedition across the Westphalian mountains. On the 13th of October he was at Nuys with thirty-two battalions and thirty-six squadrons, and still expected considerable reinforcements, but without waiting for these, on the 14th he advanced to Meurs, and with his van-guard attacked and carried Rhinberg.

*Advance of De Castries' army.*

It now became necessary for the Hereditary Prince, either to relinquish the enterprize, or by defeating De Castries, prevent his succouring the town: he therefore marched from before Wesel, about midnight of the 14th, crossed the Rhine in the morning, and on the 15th assembled a corps of twenty-one battalions and twenty squadrons, amounting to near ten thousand men, at Offenberg near Rhinberg, behind which latter place the enemy were encamped with their left extending opposite to Closter camp, and the old unnavigable canal which reaches from the Rhine to

*Hereditary Prince crosses the Rhine to attack the enemy.*

the

the Maese, immediately in their front. Having previously reconnoitered their position, he determined that night to attack their left; therefore, after leaving a small corps to watch Rhinberg, and the movements on their right, he marched at eight o'clock in one column, made a considerable circuit, arrived on Fifcher's corps posted round CLOSTER CAMP, dispersed them, passed the convent, crossed the canal (which is every where choaked up with mud and water) over a narrow bridge which the enemy had prepared for their own accommodation, and advanced along a small heath to the village of Camperbroeck and line of wood and inclosures, behind which the enemy were encamped, and who were now alarmed and prepared for their defence.

**BATTLE OF CLOSTER CAMP. Oct. 16th.**

The action began long before day, and for several hours successive attacks were made by the infantry; but the nature of the ground on which we stood, preventing our extension to either flank, it became impossible to force so superior an enemy at his center, which he could constantly support and relieve. About nine o'clock of the 16th our whole infantry totally exhausted, and having expended their ammunition, gave way and retired in disorder towards the canal, followed by the regiment of Normandie of three battalions, which advanced in perfect order and had not been engaged. In this critical situation, while the German cavalry also was retiring, the Hereditary Prince ordered Eliott's brigade of British cavalry (1st, 6th and 10th dragoons) to charge the enemy; they broke them, drove them into the wood, made near three hundred prisoners with two pieces of cannon, gave time for passing the canal, and then themselves also crossed it, protected by a brigade of British infantry which had formed the reserve, remained near Closter Camp, was now come up and lined the banks of the canal. This effort effectually checked the pursuit of the enemy; the army re-assembled some miles off on the heights of Alpen, and from thence marched towards their bridge on the Rhine. Our loss was considerable in proportion to our numbers, and fell heavy on the British troops; that of the enemy was still greater, amounting at least to three thousand men. On this oc-

casion

casion, we experienced the uncertainty which so often attends night attacks however well planned, our guides misled us; for instead of crossing the canal where we did, and thereby falling in upon the enemy's center; had we turned to the right close to the convent, and followed the road which led to Meurs, it would necessarily have brought us on the left flank of their cavalry and of their whole army. The consequences that would have flowed from such a situation, are hardly to be doubted; they must have crowned with success the gallantry and exertions of the troops, and of our commander who was here wounded.

When we arrived in the evening at the Rhine, we found that our bridge did not exist; for the waters having risen, it had become necessary to remove it at the very instant of the battle, and every exertion was now making to restore it over a river so rapid, and near one thousand feet broad. In this critical situation, exhausted with fatigue, ill supplied with ammunition, and having an imperfect communication with the rest of the army; the troops lay upon their arms that night and next day, behind an old Landwehr which here cut off a great bend of the Rhine, and to which the enemy advanced in force, and now threatened to attack us. Our bridge being ready, about two in the morning of the 18th we began to repass the Rhine, the whole had crossed it by nine, our rear from the precautions taken suffered little, nor on that or the preceding day, had we felt all the exertion that might have been expected from a superior and successful army. *Army repasses the Rhine and raises the siege.*

The siege having been raised on the loss of the battle, the artillery was sent back towards Munster; Kilmansegge with seventeen battalions and ten squadrons returned to the Dimel; and the Hereditary Prince with thirty battalions and twenty-two squadrons marched as soon as the passage of the river was compleated to Bruynen and Schermbeck, where he remained for some days to cover the march of the artillery, and on the 26th fell back to Klein Reckum. The enemy attempted no pursuit, being satisfied with having obtained their point in the relief of Wesel. In this *Camp of Klein Reckum.*

camp of Klein Reckum, and without any material occurrence happening, the troops remained till the middle of December, when they went into winter quarters in the country of Munſter.

Had this enterprize on the Lower Rhine proved ſucceſsful, it might have transferred the ſeat of war into an enemy's country, and would undoubtedly have prevented Broglie from eſtabliſhing himſelf in Heſſe; but the loſs of the battle of Campen, and the raiſing of the ſiege of Weſel, put an end to all expectation from that quarter. While this event was depending, except an incurſion which the enemy made into the country of Halberſtadt, nothing very material had paſſed between the two principal armies.

*Blockade of Gottingen.*

But Broglie now determined to maintain *Gottingen* alſo, worked hard to put it in a ſtate of defence, furniſhed it with proviſions and a large garriſon, and then drew back the corps which he had in that neighbourhood to the Weſer and the Werra. Prince Ferdinand was ſtill in hopes, that he could diſlodge the enemy from this advanced point. He therefore quitted the exhauſted poſition of the Dimel, where with difficulty he had ſubſiſted till this time, croſſed the Weſer, and on the 27th of November blockaded Gottingen; but the very late ſeaſon of the year ſoon obliged him to relinquiſh this deſign, and to leave the enemy in poſſeſſion of all their conqueſts.

Both armies now ſeparated about the 10th of December, and went into *winter* quarters—The Hereditary Prince with twenty-two battalions and eighteen ſquadrons in the country of Munſter; Sporken with twenty battalions and eighteen ſquadrons towards Duderſtadt and Hildeſheim covering Hanover; the body of the army was placed in the biſhopricks of Paderborn and Oſnabruck, and poſts on the Dimel were ſtill preſerved by the light troops.

*Winter quarters.*

As to the French, the protracted operations of the campaign had ſo exhauſted the country of Lower *Heſſe*, that Marechal de Broglie, after leaving ſtrong garriſons in Caſſel and Gottingen, could only place an inconſiderable portion of his army in it, the reſt he quartered in a very extended manner on the Frontiers of Saxony, Upper Heſſe, Fulda, and on the Mayn and Rhine.

1761.

1761.

In this situation, the armies remained for two months; and the French were employed with great industry in forming and collecting magazines for future operations—But Prince FERDINAND, who foresaw the great advantages which their present position gave them in opening the campaign, and who was in hopes that they were not immoveably fixed in their new establishments, felt the necessity of using every effort to dislodge them; and notwithstanding the shattered state of his troops, who had so long kept the field; secretly prepared and arranged every thing needful for the attempt.

*Prince Ferdinand's preparations.*

For this purpose, in the beginning of February 1761, his whole army, which embraced the circle from the borders of Saxony to those of Holland, was put in motion to unite and concentrate in the country of Hesse—The body of it crossed the Lower Dimel, and advanced upon Cassel. The Hereditary Prince on the right from the bishoprick of Munster, marched by the Upper Dimel to Fritzlar, on the enemy's communication. Sporken on the left being joined by a large corps of Prussian cavalry, gained a considerable advantage over the enemy on the frontiers of Saxony; and in pursuing his progress, cut in upon the Fulda to the very center of their cantonments. Several other small and intermediate corps, acted in conjunction and communication with these; and Munster remained secured by the garrisons of that city and Lipstadt.

*Movements of the army towards Hesse in February.*

Marechal BROGLIE thus pressed on all sides, after strengthning Cassel, was obliged to abandon the whole country, and to retire with such force as he could collect, upon Homburg, Hirschfield, Fulda, Giessen, and finally upon Francfort; where in the position of Bergen, he ordered his army to assemble, and the troops from the Lower Rhine to march thither.

*Broglie's retreat towards the Mayn.*

*General position of the army during the siege of Caſſel.*

In the mean time, the ſeveral corps of the allied army always advancing and approaching each other, puſhed and followed the enemy to the ſtrong poſition of Homburg on the Ohm; where they united on the 28th of February, and remained to cover the ſiege of *Caſſel*, which had been immediately begun under the direction of La Lippe Buckeburg—Marburg was poſſeſſed—The Hereditary Prince with a ſtrong corps was advanced towards Gieſſen—the light troops penetrated to the neighbourhood of Hanau and Dettingen—the enemy every where gave way, and ſuch had been their firſt conſternation and uncertainty, that Gottingen was for ſome hours evacuated, and the garriſon had begun its retreat.

*Advance of Broglie towards Caſſel.*

So far had this great operation proved ſucceſsful; but the debilitated ſtate of the army prevented its completion. From this reaſon, as well as from the unfavorableneſs of the ſeaſon, did the ſiege of Caſſel (on the taking of which alone the realizing of all our other advantages depended) linger on till the middle of March. At which time Marechal de Broglie, who had aſſembled a very ſuperior army in the neighbourhood of Francfort, rapidly advanced to our front, drove in the Hereditary Prince's corps with great loſs from Grunberg, and begun to extend to our left.

*Retreat of Prince Ferdinand on the Dimel.*

Prince Ferdinand, who found his nominal army unequal to the defence of the ſtrong poſition of the Ohm; that he could no longer ſubſiſt it in this advanced ſituation; that after three weeks open trenches, no great progreſs was made at Caſſel; that he could not prevent the French moving to his left, and thereby relieving it; and that all proſpect of ſucceſs from perſeverance was now at an end; ordered the ſiege to be raiſed. After remaining with his army au Bivouac for three days behind the Ohm, in order to check the progreſs of the enemy; he retired on the 23d by Newſtadt, Treyſa, Wildungen, Fritzlar, Wolfshagen; and on the laſt day of March croſſed the Dimel, at the ſame time that the raiſing of the ſiege was perfectly compleated without any conſiderable loſs.

Broglie,

Broglie, who had gained some advantages in this retreat, pressed it no farther than Fritzlar, and himself entered Caffel on the 29th, satisfied with the credit of having thus relieved it. On the whole his conduct was judicious, showed spirit and resource, and was crowned with success: The commandant Comte de Broglie, also acquired reputation from his defence. After this hard service, the two armies immediately went into quarters of refreshment—That of Prince Ferdinand (leaving the Dimel guarded) in the countries of Munster, Paderborn, and Osnabruck—That of Broglie returned on the Mayn and the Rhine.

*Cantonment of the armies, 1st of April.*

Was the whole of this *winter* campaign fully analised, it might perhaps be found not inferior in the conduct to the famous irruption of TURENNE into Alsace, when he forced the German army to repass the Rhine—The scene of operations was here in a more wide and extended country—The previous distribution of quarters, was made with a view to the subsequent event, and without alarming the enemy—The movements of the troops from the circumference of so great a circle, were connected, just, and wonderfully combined in their effect of dislodging the enemy, who surprized and unprepared, could no where assemble in any of the strong positions which Hessia offers—The arrangements for the siege and for subsistence in that season, were made with skill; nor did the army ever experience want, although so long acting in a very advanced situation, and in a country supposed to be exhausted—Much resource was certainly found from the French magazines which fell into our possession—The service was severe; for during the months of February and March the movements of the troops were unremitting, sometimes made in cantonments, but often under canvas or the open air—When perseverance had done its utmost, and it became necessary to retire, although some of the rear corps received checks, yet the siege was raised, and the

*Remarks on this expedition.*

the army fell back without any brilliant advantage on the side of a superior enemy.

<small>Great objects in view.</small>
The *object* of this expedition was great and worthy of trial; and had it been succefsful, might have given a final turn to the war. If Caffel fell, Gottingen followed of courfe, and twelve thoufand men of the beft troops of France were prifoners; nor would the advantages have been confined to the repoffeffion of Heffe—The failure alone arofe from the weak ftate of the army, which proved unequal to the execution of this undertaking; and from fome peculiar accidents which could not be forefeen, and turned out unfavourable. The other calculations of the general were juft; for three weeks of open and undifturbed trenches, ought in any other cafe to have decided the fate of Caffel, and the fortune of the war.

The deftruction of the French magazines in Heffe, and the other confequences of the SPRING EXPEDITION, which had extended its effects as far as the Mayn, retarded the opening of the campaign of 1761 on that fide, and determined the efforts of the enemy towards the countries of Paderborn and Munfter, where the taking of LIPSTADT became their principal object. In the beginning of June, the army on the Lower Rhine, commanded by Marechal de Soubife, was in motion to affemble being greatly reinforced from France, although Broglie's army had not begun to collect in Heffe, the country having been entirely exhaufted, and the green forage as yet affording no fubfiftence.

<small>Situation of the Britifh army the beginning of June.</small>
This obliged Prince Ferdinand to encamp a corps under Lieut. Gen. Conway at Soeft, on the 12th of June, from whence Lieut. Gen. Howard with four battalions and three fquadrons, was detached on the 15th to Ham to fecure that place, and to communicate with the Hereditary Prince, who about this time formed his camp at Nottlen, and covered the country of Munfter—The reft of the army remained cantoned in the bifhopricks of Paderborn and

and Munster, ready to march. Hanover was protected by light troops and a small corps under Luckner.

Soubise's army, consisting of one hundred and nine battalions and one hundred and eight squadrons, crossed the Rhine at Wesel the 14th and encamped on the 18th at Dormundt—This occasioned the Hereditary Prince to quit his position at Nottlen on the 19th and approach the Lippe: he encamped on the heights opposite Ham on the 21st—The body of the army marched from their cantonments, and on the 19th encamped at Paderborn, as did Sporken's corps the same day on the Dimel, near Warburg.—The establishment of the armies now consisted of, but the French were more than proportionally complete.

*Soubise's army.*

*Camp at Paderborn, June 19th.*

French.
- Infantry - 105,000
- Cavalry - 26,000
- Light troops 10,000
- —————
- 141,000

- 109 battalions } Soubise's army
- 108 squadrons
- 88 battalions } Broglie's army
- 89 squadrons

Including their light troops, communications, and garrisons in Germany.

British.
- Infantry - 68,500
- Cavalry - 15,160
- lt. troops { Cavalry 3,000
-            { Infantry 5,000
- —————
- 91,160

- 97 regular battalions, including grenadiers.
- 5 battalions of Hanoverian militia.
- a large body Chasseurs and irregular infantry.
- 84 squadrons of cavalry.
- 21 squadrons of Hussars.
- 3 squadrons British legion.
- mounted Chasseurs, a considerable body.

Carriages of all kinds, about - 5,000 } besides country carriages,
Horses of all kinds, about - 58,000 } impressed and employed.
The daily consumption of forage, little short of 70,000 rations.

At this time, the situation of the different corps which composed the BRITISH army, was as follows, viz.

Situation of the army the middle of June.

|  | Bat. | Squad. | Light Troops. |
|---|---|---|---|
| Luckner upon the Weser with | 4 | 4 | Hanoverian chasseurs, Brunswick hussars, Stockhausen's corps. |
| Sporken upon the Dimel | 15 | 16 | Hessian chasseurs. |
| Army at Paderborn | 28 | 14 | |
| Wangenheim near Buren | 7 | 4 | 1 battalion and 1 squadron of the legion, 4 squadrons Brunswick hussars. |
| Conway at Soest | 8 | 7 | 5 squadrons Prussian hussars. |
| Howard at Ham | 4 | 3 | |
| Hereditary Prince at Nottlen | 26 | 24 | 4 squadrons Hessian Hussars, Scheiters, Trembachs, 2 battalions of the legion. |
| Munster, Lipstadt, Hameln Rinteln, Minden garrisoned | 5 | 2 | Exclusive of 5 battalions Hanoverian milice, and 2 battalions 2 squadrons of the legion. |
|  | 97 | 84 | |

A. Camp at Stadt Geseke.

In order to stop the further operations of the enemy, the army marched on the 21st of June from Paderborn to Stadt Geseke in six columns.—On the 22d the Hereditary Prince's advanced posts at Lunen, Unna, and Kamen, were forced in by Conde's reserve, which encamped the 23d at Unna; on which day our army moved forward from Stadt to Alt Geseke.

B. Camp at Alt Geseke.

Plates { 18. 19.

C. Camp at Soest.

On the 24th at one in the morning the army marched in six columns to Soest where it encamped, and was joined by Conway's corps—Soubise's army moved forward this day to Unna. On the 25th Lord Granby with six battalions and six squadrons was advanced to Wippringhusen; Howard marched to Kirk Denkeren; and the Hereditary Prince crossed the Lippe and encamped at Ham. On the 26th Howard with his corps joined Lord Granby, and the Hereditary Prince occupied the heights of Kirch Denkeren, leaving Kilmansegge with eight battalions, and four squadrons to cover Ham. On the 27th at night the castle of Werle garrisoned by a detachment of our infantry was attacked, but the enemy were repulsed. In the evening Lord Granby's corps marched, and lay under arms behind Rhunne.

On

On the 28th at one in the morning the army marched in six co-    **D.**
lumns, and encamped at Werle, where the Hereditary Prince joined
on the right—The enemy abandoned Neheim on the Roer, (which   Camp at
we took poffeffion of,) and drew in all their advanced pofts nearer to   Werle.
the ftrong camp of Unna, which they now occupied.

On the 29th at four in the morning, the army marched from
the right in nine columns.

|  |  |  | Batts. | Squads. |  |
|---|---|---|---|---|---|
| H. Prince | {1ſt} {2d} |  | 18 | 22 |  |
| Army. | 3d | — | 6 | 7 |  |
|  | 4th | — | 1 | 0 | Britiſh artillery. |
|  | 5th | — | 6 | 7 |  |
|  | 6th | — | 8 | 6 |  |
|  | 7th | — | 1 | 0 | Hanoverian artil. |
|  | 8th | — | 9 | 5 |  |
| Ld. Granby. | 9th | — | 10 | 9 |  |
|  |  |  | 59 | 56 |  |

Diſpoſition of march to Lunderen.

The light troops and piquets, formed the advanced guards of the
columns, and the whole was in expectation, and readineſs to meet
the enemy, and in an inſtant to form in line of battle; but Soubiſe    **E.**
did not move from his camp of Unna—Condé's reſerve which had
occupied the heights of Oſt Buren, was therefore driven in upon   Camp at Lunderen.
the enemy's army, and the camp eſtabliſhed at Lunderen. Beck-
with's brigade took poſt at Cloſter Sheide, and the next day joined on
the left of Lord Granby.

On the 30th in the morning Kilmanſegge's corps from Ham,
appeared on the heights of Kamen—The enemy retired the right
of their camp, and cannonaded ours with little effect—This and
the following day, Prince Ferdinand was employed in reconnoitring

their

**H.**
Soubife's camp at Unna.

their fituation which was found exceedingly ftrong: The right flank covered by the river Roer, beyond which lies the mountainous, woody and barren dutchy of Weftphalia; the left by Unna, feveral marfhy rivulets, and a deep country betwixt, and the Lippe; the front was defended by a line of redoubts and entrenchments thrown up with incredible labour, and before which ran a deep ravine, wooded on each fide, and paffable only in three places.

In this fituation it appeared to be Soubife's intention to wait the approach of Broglie, to cover the march of the heavy artillery from Wefel, and to make every neceffary preparation for the fiege of Lipftadt, which was to be the great object of their joint operations, but for which purpofe a fure communication with the Lower Rhine was abfolutely neceffary, as from thence the principal fupplies for both armies muft be drawn.

The irrefolution and timidity apparent in Soubife's conduct confirmed Prince Ferdinand in his defign to attempt fome coup d'eclat before the enemy could be fupported by Broglie who was now within five marches of us, and who affembling his army of eighty-eight battalions and eighty-nine fquadrons, had croffed the Dimel on the 29th and obliged Sporken to retire upon Blomberg.—Wangenheim alfo fell back towards Lipftadt July 2d.

Preparations to attack Soubife.

There was now no time to be loft—The enemy was unattackable in front and on either flank, their rear alone might be acceffible, and it was effential to diflodge them from this ftrong poft, before the junction of their armies was effected near this point, and that their combined operations fhould oblige Prince Ferdinand to pafs the Lippe—The troops were therefore fupplied with bread for feveral days, their waggons returned to Lipftadt, and as foon as it was dark the tents were ftruck, and fent with the baggage to Ham. At eleven at night of the 5th of July the army marched from, and to the right in five columns as encamped, viz.

The 1st or left column, composed of the infantry of the 1st line.  
  2d   —    —   infantry of the 2d line.  
  3d   —    —   line of cavalry.  
  4th⎫  
  5th⎭  —    —   artillery.

*Disposition of march to Dortmundt.*

The Hereditary Prince's corps which was joined by Kilmansegge at Kamen, formed the van-guard of the three left columns of the army, and Lord Granby's corps formed the rear-guard in like order.

About day break the rear-guard had passed Lunderen, and got into the inclosed country, without any considerable annoyance from the enemy; but an uncommon heavy rain, which fell in the night rendered the routes almost impassable, and prevented the columns arriving on the heights of Kamen before mid-day; in this situation we were opposite the flank of Soubise, who as yet had made no movement, imagining we were going off towards Ham. About two leagues beyond Kamen, the columns halted for some time, with a view of directing them to the open ground within three miles of the enemy's rear, which we had now gained, and who by this time were exceedingly alarmed, and had struck their camp. But the difficulties of the country, and the absence of the artillery, determined that the columns should resume their march upon Dortmundt, where the front arrived about eight at night. The rear however and the piquets, which flanked the march of the columns on the left, and repulsed an attack of the enemy on a temporary post of ours established at Wasser Kurle, did not get up till ten next morning, nor was all the artillery arrived before mid-day of the 3d.—This march was lengthened at least twelve hours by the unlucky rains which had fallen, and prevented our immediately advancing on the enemy.

*March to Dortmundt, and difficulties.*

The tents were now ordered up, and the army encamped at Dortmundt on the 3d; the Hereditary Prince being advanced two miles in front, and Lord Granby's corps, which had remained the night of the 2d near Meidel, encamped also at Lanstrup. The enemy, who on the 3d had altered their encampment and now faced

*F. Camp at Dortmundt.*

towards us, about six in the evening struck their tents and marched upon Werle, not venturing to await us in their position of Unna, which presented a camp sufficiently strong to the Dortmundt Front, and to dislodge them from which would have much puzzled us, had they taken such a resolution. Perhaps the interruption of their communication, and the difficulty of subsistence determined this retreat; whereas our communication with Luynen and Ham was still preserved.

On the 4th at one in the morning, the army now consisting of sixty-nine battalions and fifty-five squadrons, advanced in six columns from the right, preceded by the vanguards.

Disposition of march to Unna and Hemerden.

Lord Granby {5——5          8—— 8——10 Batts.} Hered. Prince.
           {6——3    .   .  8—— 8—— 8 Squads.}

Army — { Hanoverian  1——5——8——8——10—— 1 Batts.  } Army
        {            Center  Center      British Artill. }
        {     5        6              7      7 Squads. }

The noble routes which the enemy had already opened in advancing from Dortmundt, facilitated this march. About six we arrived in their camp of Unna, and saw their rear-guard on our former ground at Lunderen: the armies having thus exactly changed situations. The Hereditary Prince's corps pushed forward, drove the enemy from hollow to hollow, and cannonaded them briskly, till they formed under the protection of the strong Landwehr, which extends from Buderick to the Roer. The number of deep hollow ways, with which this whole open and fertile country from Paderborn to Dortmundt is cut, and which are not apparent till on the very brink of them, prevented our closing with the enemy, and enabled them thus frequently to make a stand.

Soubise's

Soubife's army, now on their march to Soeft, finding their rearguard so warmly pressed, came back and formed between the castle of Werle (which we still possessed) and the woods. The heads of our columns advanced close to the enemy, but finding them in possession of the heights over the Roer, which we had endeavoured to gain, strongly posted, and sustained by their whole army, we fell back at four o'clock and encamped at Hemerden. Camp at Hemerden. G.

At five o'clock on the morning of the 5th, the army marched to attack the enemy, in the same number of columns and disposition of march as yesterday: the Hereditary Prince and Kilmansegge to attack on the right next the Roer, and lord Granby on the left at Budrick. The heads of the columns had advanced and halted within cannon shot, and a few were fired on each side, when the Duke sent for the lieutenant generals, acquainted them that he found the enemy too strongly posted to be attacked, that they were formed behind the principal Landwehr, which he thought we had gained (for we were in possession of one) and that he was therefore obliged to desist from his original intention. Attack of the enemy postponed.

This manœuvre however appears to have been a feint calculated to preserve the idea of superiority, and to try the countenance of the enemy, who might have been induced to abandon their present position, as they had before done that of Unna.—From the operations of yesterday the Duke must have known their exact situation, which was now exceedingly strong, as they had worked hard all night, and converted the Landwehr into an intrenchment which covered their whole front. Reasons.

About nine the army returned to their ground and encamped—The enemy still remained au Bivouac, nor had they any regular camp formed since the 29th of June; such was the panic that had struck them, from the bold, rapid, and unexpected movements of the British army. In the evening Beckwith with four battalions, three squadrons, Eliott's, and six squadrons of Hussars, took post at Borg and Budberg, on the enemy's right flank. Situation of the enemy.

The

The enemy still continued on the 6th fortifying their camp; but as Broglie was now advancing from Paderborn, and could soon act in conjunction with Soubise's army, it became necessary to quit our present exposed situation and approach the Lippe, in order to preserve our communication with Ham, from whence our supplies were drawn.——On the 7th therefore at two in the morning, the army marched with the greatest circumspection, by the left, in five columns, to the camp of Hillbeké.

<small>March to the camp of Hilbeké.
L.</small>

1st or right column, composed of the infantry of the 1st line.
2d — — — infantry of the 2d line.
3d — — — cavalry.
4th 5th — — — artillery.

Lord Granby's corps formed the advanced guard of the three first columns: the Hereditary Prince's corps formed the rear-guard, and the piquets with cannon flanked the march. Sheidingen was occupied by Campbell's Highlanders, Eliott's, and the Hussars.

<small>March of Soubise to Soest.</small>

When day broke, we found that the enemy also had marched off in the night in the greatest silence, and were now past Rhunné on their way to Soest, to effect a junction with Broglie's army which was advancing upon Erwite, after leaving Prince Zavier's reserve in the neighbourhood of Paderborn. In this manner were the two French armies drove in upon each other, and their direct intercourse with the Lower Rhine thoroughly interrupted.

The British Army was now situated as follows.

<small>Situation of the British army.</small>

Freytag with his chasseurs in the woods upon the Weser, to interrupt the enemy's communications with Gottingen, and cover Hanover. Luckner with his small corps watched Prince Zavier—Wangenheim had retired on the 5th under the cannon of Lipstadt.—Sporken was advancing from Blomberg to Hertzfeld, where he arrived the 8th.—The army and Hereditary Prince occupied the heights of Lunderen and Wambeln—Ham was in the rear of the army, and thereby the communications with Lipstadt and Munster on the

the other side of the Lippe were kept open, as from them our principal supplies were drawn.

On the 8th, Broglie arrived at Erwite—Sporken also encamped at Hertzfeld; and Howard was detached with four battalions, three squadrons of Eliott's, and five of Hussars to cover Ham— The French armies thus united, could evidently not long remain inactive. Far removed from their magazines, and their communications precarious and of great extent, they were under the necessity of giving up the first and great object of the campaign, or of forcing Prince Ferdinand to pass the Lippe. *Junction of the French armies.*

*K.*

From the position of Broglie's army (which formed the right of the enemy) near the Lippe, from the enterprizing spirit of that general, and from the strong reconnoitring parties that were pushed daily along the great road from Lipstadt towards Ham; it became necessary to cross the Aas, extend our left, and occupy the heights of Kirch Denkeren, which was done between the 10th and 14th, by eighteen battalions and fourteen squadrons under Lord Granby and Wutgenau; the right of the army being proportionably thinned, but still occupying the whole of the former position. Fleches, redoubts, and various field works were constructed at different points, along the whole front; bridges were laid across the Aas; heavy baggage was sent to Ham; every thing was kept in constant readiness and expectation of an attack; and the advanced posts of Budrick, Werle, Sheidengen, Meyericke, Welfert, Madena, Hufnehel, Nardel, Hiltrup, Heyntrup, were occupied by strong detachments. *Disposition made to receive them.*

On the 13th, Soubise advanced and encamped between Closter Paradys and Rhunné; and on the 14th, all our advanced posts on the right, beyond the Saltzbach, were drove in. Howard's brigade was therefore ordered to recross the Aas, and Wutgenau's corps to move from the left to Kirch Denkeren, as our right seemed to be immediately threatned. *Advance of Soubise's army.*

*P.*

But

But this proved to be only a feint—For on the 15th, Broglie marched early in the morning from his camp near Erwite in four columns, advanced between the Aas and the Lippe, drove in our light troops and out posts, followed them, and before sunset arrived on the heights of Denkeren in front of Lord Granby, whose camp he cannonaded while the troops were forming; he then pushed along the Ham road to the heath in the rear of FELLINGHUSEN; but was soon obliged to retire from the resistance made in the wood and castle of Fellinghusen, by a battalion of the legion and the Highlanders, and by the timely arrival of Wutgenau's corps from the right.

*Attack of Broglie on the 15th at* FELLINGHUSEN.

L.

In this manner the night closed, the armies within musquet shot, occasionally keeping up a scattered fire, and separated on our left by a single inclosure and a Landwehr, of which we were in possession. Soubise also, who had marched about mid-day, was formed in different corps behind the Saltzbach, in front of Conway and the Hereditary Prince, but made no attack in the course of the evening.

During the night, the Duke determined on reinforcing his left; different corps marched; the right was anew arranged; and before day light, the position of the armies was as follows.

*Position of the armies on the 16th.*

### BRITISH.

| | Bat. | Sq. | Huf |
|---|---|---|---|
| The Hereditary Prince behind the Saltzbach, between Hillbeck and Ellingsen, with | 26 | 24 | 6 |
| Conway, between Ellingsen and Hohenover, with | 8 | 7 | 0 |
| The body of the army on the heights of Kirch Denkeren, and between the Aas and the Lippe | 35 | 27 | 6 |

### FRENCH.

| | Bat. | Sq. |
|---|---|---|
| Duvoyer near Unna, with | 12 | 14 |
| Dumesnil near Budrick | 16 | 30 |

Soubise with the rest of his army along the Saltzbach, between Werle and Closter Welfert.

Broglie with his army in different lines and columns, between the Aas and the Lippe.

At day break, Broglie renewed his attack, extending from the heights of Denkeren by Fellinghufen to the Lippe, along an extent of about a mile; but such were the difficulties and intricacy of the ground, and the gallant refiftance of our troops, that he made no impreffion after repeated attempts; and at laft about ten in the morning (on the arrival of fix battalions and fix fquadrons detached from Sporken's corps, who had croffed the Lippe at Hus Haren where bridges were eftablifhed) the enemy began to give way; nine pieces of cannon, and the four battalions of Rougé were taken with many others prifoners; the enemy were preffed and followed to Heintrup, from whence the nature of the country favouring them, they retired upon Hoveftadt and Oeftinghufen, without farther purfuit.

*Attack of Broglie.*

*R.*

While fuch were the tranfactions on the left, Soubife meeting with many impediments in advancing, had attacked but faintly on the right, and had never been able to crofs the Saltzbach in any confiderable force, nor to pafs through the village of Sheidingen; being prevented by the refiftance made in an old redoubt, gallantly defended by the piquets of the army. On Broglie's defeat, which he was immediately apprized of, he drew off, and retired to his camp at Paradys, from whence next day he fell back to Soeft. After the battle, our tents and baggage were ordered up from Ham; the army was directed to repoffefs, and encamp on the fame ground it occupied on the 15th; and additional works were alfo begun on the Dinkerberg.

*Attack of Soubife.*

*R.*

Although the pofition at FELLINGHUSEN was very extenfive, being feven miles from right to left; yet were the communications between the different parts of the army fo well eftablifhed, the front fo well covered by natural and artificial obftacles, rendering the approach through an intricate country exceedingly difficult, that the columns of the enemy could not affift each other, or act in conjunction—Their efforts were reduced to certain points of attack, thofe points were ftrengthned, the fituation of the allied army was every where commanding; a retreat

*Nature of the pofition of Fellinghufen.*

was secured upon Ham a few miles in the rear, and the whole was so judiciously posted and supported, that their attempts were repulsed with ease—Nor was there ever any considerable impression made after the evening of the 15th, which was indeed a kind of surprise to which every army must be liable in a country like this, flat and inclosed; where a superior enemy lying so near, could assemble unseen behind his advanced posts, and at once move forward in force to the attack.

Original intention of the enemy,

The enemy sensible of the difficulties they had to encounter, had agreed to have employed the 16th in forcing in our out posts, establishing themselves along our whole front, gaining our right flank and rear, opening communications between their several columns; and on the 17th, to have made a general and extended attack with their superior numbers. But the ardour and confidence of Broglie broke through this plan, before it could be put in execution in its several parts by Soubise, hastned the subsequent defeat of the two armies, and occasioned that want of concert, which was so apparent in the operations of that day, upon which Prince Ferdinand gained so memorable a victory.

not executed.

Importance of this position.

It is remarkable that this position of the British army (at least the right of it) was the same occupied by Contades in the end of 1758, when he detached a considerable part of his army to reinforce Soubise in Hesse, who was thereby enabled to defeat Oberg near Cassel. Prince Ferdinand at that time marched from Munster by Lipstadt, Soest, and Werle, to attack Contades; but was obliged to return without venturing to effect his purpose, finding him too strongly posted. The strength of this position seemed by degrees only to have been discovered, for several days the Dinkerberg was unoccupied; and had the enemy on the 9th, 10th or 11th, made the same rapid march which they did on the 15th, and gained possession of it; Prince Ferdinand would possibly have been obliged to fall back on Ham, and to pass the Lippe.

Such

Such was the event of the battle of FELLINGHUSEN; which effectually deranged the hopes and the schemes of an enemy, greatly superior in numbers, although their loss in men from local circumstances was not very considerable—The jealousy which subsisted between the two generals, now openly broke out; each supported by his party, accused the other of precipitancy or delay; the court was appealed to, and the armies remained inactive till further orders were received. *Consequences of the battle of Fellinghusen.*

The French armies having now only a mountainous intercourse with the Lower Rhine, and drawing their supplies of bread from Cassel and their scanty magazines in Hesse, could not long remain together—Luckner also with a small corps, had all this time hovered about their rear, in the country of Paderborn, and very much disturbed their communications. On the 19th, Broglie fell back to Erwite, and Soubise to Soest; our advanced posts were pushed forward, and nothing considerable occurred for several days. *Difficulties of the enemy.*

On the 25th, the French armies which had with difficulty subsisted, now separated—Broglie, who had undertaken for the future success of the campaign, being reinforced from the other army with thirty-six battalions and fifty squadrons, marched back on Paderborn; and Soubise crossed the Roer, entered the mountains of Westphalia, marched by Arensberg, and with the remainder of his troops much shattered, again came out into the low country between Wesel and Dortmundt, where about six weeks before in a flourishing state he had begun the campaign—The household troops unequal to such rude service, were sent into cantonment, and did not again appear in the field. *Their armies separate and retire.*

A new scene of operations, and in another country now took place—The Hereditary Prince with his corps was left near Werle to watch the motions of Soubise's army, and prevent his too soon issuing from the mountains; the army marched on the 27th, and advanced successively, *Advance of the British army.*

To Borglen in six columns ⎱ By the spacious and
To Erwite in six columns ⎬ noble routes opened
To Stormede in ten columns ⎰ by the enemy;

and to Buren in five columns, where it encamped the 30th, having been previously joined by Sporken and Wangenheim; and then amounting to sixty-one battalions, fifty-five squadrons, thirteen squadrons hussars, besides irregular infantry.

Broglie's object was now to attempt the Siege of HAMELN. For which purpose he placed his army in different corps, behind the Dimel at Stadberg, and the hills and woods of Paderborn, and made every preparation for forwarding his artillery from Cassel—But Prince Ferdinand seeing the country round Lipstadt thoroughly exhausted, and that there was no probability of a return of the enemy to that place; marched on the 10th of August by his left, successively to Delbruck, Stuchenbroeck, Detmold, and on the 13th to the strong camp of Blomberg, which effectually covered Hameln—Broglie prevented in every other object, was now obliged to cross the Weser on the 17th and 18th at Hoxter, and endeavour to subsist in the country of Hanover; to this determination, the rapid march of the Hereditary Prince to Buren on the 12th (after leaving a small force to watch Soubise) and his operations from thence on the rear of Broglie's army, did not a little contribute.

*Broglie obliged to cross the Weser.*

---

During the rest of the campaign, Soubise much weakned by the detachment he had made to the other army, was not able to undertake any thing substantial, either against Munster or Lipstadt, although his light troops occasionally overran and plundered Friesland, and the bishopricks—Broglie's army was extended in different corps from Eimbeck, by Gottingen, Munden, Cassel, Fritzlar, Ziegenhayn, Marburg, to cover their communication with the Upper Rhine, and strengthen the posts they already occupied—The various movements of the British army kept them

*Progress of the campaign.*

in constant alarm; several inconsiderable actions happened; and Prince Ferdinand held a central situation near the Dimel ready to move on either army, should they attempt any operation of importance.

In the month of October, Broglie having collected supplies, and being enabled to give up his communications for a time, made a last effort on the country of Hanover, but was repulsed with loss in his attack on Brunswick—This drew Prince Ferdinand from the Dimel towards Hameln, and the beginning of November he crossed the Weser at that place, marched upon Eimbeck, and by a series of masterly manœuvres, dislodged Broglie who it was thought had an intention to establish himself there, and forced him back on Gottingen. The severity of the weather now obliged the armies to separate, and go into winter quarters, each remaining in possession of the same country with which they had begun the campaign; but much to the honour of Prince Ferdinand, who had in every instance resisted the great efforts of France, and baffled the attempts of such superior enemies.

*End of the campaign.*

The quarters of our army extended from the frontiers of Holland to those of Saxony; and their front was covered by the advanced posts of Coesfeldt, Dulmen, Halteren, Ham, Neuhans, Driburg, Brakel, Hoxter, Dassel, Eimbeck, and Osterode, in the Hartz Forest. The French extended from Muhlhausen, through Hesse and the Veteravie to the Rhine, and behind it; having their advanced posts at Muhlhausen, Heiligenstadt, Gottingen, Munden, Cassel, Waldeck, Corbach, Winterberg, Arensberg, Iserlohn, Hattingen and Doesburg on the Rhine.

## 1762.

The winter passed without any attempt of consequence being made on either side, until the middle of April, when the Hereditary

ditary Prince assembling a small corps at Ham and Lipstadt, marched upon Arensberg, one of the enemy's most advanced posts; it surrendered on the 19th, was dismantled, and the troops returned to their quarters. This year neither of the armies were so numerous, or so well recruited, as in the last campaign; but the superiority of the French still enabled them to expect conquest, and as usual they prepared to form two armies. One of one hundred thousand men in Hesse, under the command of the Marechals D'ETREES and SOUBISE, and another of twenty-five thousand on the Lower Rhine, under the Prince of CONDE.

*Winter operations.*

About the middle of May, our quarters were contracted as a preparatory measure. On the 4th of June, the infantry and artillery encamped in front of Blomberg; the cavalry was cantoned in the rear, where they remained till the 17th, when they also encamped; and from the 8th, Lord Granby's reserve of eleven battalions and nine squadrons had been advanced to Brakel—Major Gen. Retz with four battalions, was near Paderborn; Luckner with six battalions and four squadrons, was near Eimbeck; and on the 19th, the Hereditary PRINCE with his corps of twenty battalions and twenty-three squadrons, encamped at Dulmen—The Chasseurs were in the Hartz, and the other light troops were divided among the several parts of the army. The enemy also was every where in motion to take the field.

*Assembly of the army.*

On the 18th the body of the army of forty-six battalions and forty-eight squadrons marched to Brakel—On the 19th Lord Granby moved to Peckelsheim, and the light troops extended along the Dimel—On the 20th Lord Granby encamped at Warburg and the army at Borgholtz—On the 21st it occupied the camp of Buhne, the castle of Sabbaborg beyond the Dimel was also possessed, and the enemy began to assemble near Cassel.

On the 22d the French moved forward to the camp of Grebenstein in front of WILHELMSTALL, having their flanks covered with detached corps, but their out posts not sufficiently advanced, and pushed on to the bank of the Dimel which they were entitled to possess—PRINCE FERDINAND perfectly acquainted with the nature of their position determined to attack them in it, and as a preparatory step now occupied several posts on both sides to secure the passages of the Dimel—On the 21st and 22d his light troops scoured the country beyond it, to gain more accurate intelligence of the enemy's situation, and he ordered Luckner, after leaving a small detachment near Eimbeck, to march so as to arrive on the 23d at night in the neighbourhood of the Sabbaborg.

Finding that the Marshals had assembled in a confined position, and had not begun to extend corps to their flanks; on the 24th our whole army of sixty-seven battalions and sixty-one squadrons, besides light troops, was put in motion, and early in the morning passed the Dimel in ten columns in various places, the advanced posts which we occupied concealed the knowledge of this movement from the enemy.—Sporken and Luckner were destined to turn and attack their right, Lord Granby (who had a great detour to make) their left, and the main army their center—The enemy astonished and unprepared to make head against so many combined efforts struck their camp, and began to retire in great confusion on Cassel, this their right and center effected without any considerable loss and before they could be joined by our line, being covered by Stainville's reserve of their left, which also would have escaped; had not Lord Granby, foreseeing such event, moved briskly on to the rear of the enemy near WILHELMSTALL, where intercepting them a very brisk action took place, which ended in the entire defeat of that corps, and the capture of many prisoners—The French were pursued and retired under the cannon of Cassel, their loss amounted to about six thousand men: our army suffered little, and encamped on the field of battle behind Wilhelmstall.

BATTLE OF WILHELMSTALL.
June 24th.

It is singular and here worthy of observation, that Prince Ferdinand during three successive campaigns had constantly in view the attacking of the enemy whenever he should find them in this position of Grebenstein and Immenhusen, and the dispositions he made for that purpose were nearly the same all the three years—He intended on the 15th of September 1760 to have marched from Warburg

burg upon Marechal de Broglie then encamped on the heights of Immenhufen, but the Marechal fell back to Caffel on the 13th and thereby avoided an action.—He croffed the Dimel in September 1761 to attack a large corps encamped under Stainville on the fame heights, who alfo was enabled to regain Caffel, by the troops deftined to turn his left, not making a fufficient circuit.—Thefe two rehearfals undoubtedly taught Prince Ferdinand on the 24th of June fo to combine his extenfive movements, that on the very opening of the campaign, he furprized and defeated a fuperior enemy who meditated fieges and conqueft, and he then gained a victory which in its confequences obliged them to abandon Hanover and Heffe, and to retire on the Mayn.

On the 25th the French army paffed the Fulda at Caffel and encamped behind it, the Saxon corps was on their right between Munden and Gottingen, Rochambeau with four battalions and eight fquadrons was on their left at Homburg, and they alfo occupied feveral inconfiderable pofts on the Werra, Eder, and Fulda, all which covered their communication with Upper Heffe, and the Mayn.—Prince Ferdinand now began to extend to his right, Lord Frederick Cavendifh with the light troops, on rhe 26th moved towards Fritzlar which the enemy abandoned, as well as Felzberg, and Gudenfberg on the 29th—On the 1ft of July, Lord Granby and Lord Frederick marched to attack Rochambeau, who neverthelefs with fome lofs effected his retreat upon Ziegenhayn: the light troops took this opportunity of extending towards Hirfchfeldt, but returned, when Lord Granby on the 2d fell back upon Fritzlar in confequence of the enemy's movements to their left.—On the 4th Rochambeau reinforced again, took poft at Homburg, and on the 7th advanced to the heights of Wavern, where he remained till the 12th, when he fell back anew upon Ziegenhayn, and from thence on the 15th to the Ohm, to avoid the attack of Luckner, who had marched for that purpofe.

The caftle of Waldeck having been taken on the 12th by Lt. Gen. Conway, and the various movements made to the right of the Britifh army, alarming the enemy much, for their communication with the

Mayn;

*Movements of the armies.*

Mayn; they extended to their left, recalled their troops from beyond the Werra, and on the 15th croſſed the Fulda with a very large corps, which was placed on the Heiligenberg, and in the woods in front of Melſungen.—The viſible movements of the Britiſh army ſtill tending to fix the attention of the Marechals to their left, and to draw their force towards that hand; Prince Ferdinand, who had ſecretly made other arrangements for the purpoſe, determined to diſlodge the Saxon corps now poſted at LUTERNBERG, and forming the right of their army. On the 23d, notwithſtanding the difficulties of ſituation, the Fulda was paſſed in ſeveral places, and the enemy were attacked in front by Lieut. Gen. Zaſtrow, at the ſame time that a corps which had marched from Uſlar paſſed the Werra above Munden, and fell upon their rear; twelve hundred priſoners and thirteen pieces of cannon were taken. After this ſucceſs our troops returned to their former poſitions, and the Saxons again encamped at Luternberg. Affair of LUTERNBERG.

On the 24th and 25th Prince Ferdinand made ſtill farther movements to his right, which ſo alarmed the Marechals, that on the night of the 25th they abandoned their ſtrong poſition at the Heiligenberg, repaſſed the Fulda, and thereby enabled Lord Granby's reſerve to occupy the heights above Melſungen, and our light troops to extend quite to the city of Fulda, of which we took poſſeſſion. From this time to the 6th or 7th of Auguſt, both armies were employed in obſerving each other, and no conſiderable changes of ſituation were made: the river ſeparated them—the French extended from Munden behind the Fulda to Spangenberg, and Stainville with a conſiderable corps was towards Hirſchfeldt. The allies extended in ſeveral bodies from Gieſmar on the Dimel to Melſungen, having Luckner on their right near Rothenburg, and a ſmall corps on their left near Uſlar, deſtined to cover Hoxter, ſecure the navigation of the Weſer, and watch the garriſon of Gottingen. General ſituation of the armies.

By this time alſo the two armies from the Lower Rhine were arrived in Upper Heſſe; for the battle of Wilhelmſtall having entirely deranged the projected and combined operations of the enemy, which were probably the ſieges of Munſter and Lipſtadt: the two Marechals found it neceſſary to call the Prince of Condé to their aſſiſtance, who had assembled at Wesel on the 25th of June, and had begun his operations by advancing on the 28th to Coesfeldt. On the 17th of July he crossed the Lippe at Halteren, and marching in four divisions by Essen, Dusseldorff, Siegen, &c. was on the 4th of August at Herborn, and on the 7th at Grunberg, from whence he communicated with Stainville. These movements were accompanied by the Hereditary Prince's corps, who taking the other side of the Westphalian mountains, crossed the Lippe at Ham on the 21st of July, and marching by Stadberg, Corbach, and Frankenberg, was on the 1st of August at Wetter, on the 7th near Kirchain, and on the 8th behind Homburg on the Ohm, not far distant from Condé.

It was evident that the Marechals having a very uncertain and circuitous communication with the Mayn, and having nearly consumed the forage in the narrow country they possessed about Cassel, must soon be obliged however unwillingly to abandon their present position, and leave that city to it's own defence. But Prince Ferdinand desirous to hasten an event, which was to realize to him the advantages acquired by the battle of Wilhelmstall, made preparations for a general and combined attack along the whole French front. On the 9th this took place at several points, but chiefly on their left, which was turned by Wangeheim (detached by Lord Granby) and Luckner with a small corps, who crossing the Fulda, took post at Spangenberg; while Lord Granby with his main body cannonaded Melsungen, and while Prince Frederick of Brunswick with some battalions and squadrons was even at Eschewegé on the Werra, in the rear of their army.—Notwithstanding these measures, the strength of the enemy's position was such, that no material impression could be made, nor would they relinquish it: On the 10th therefore, the army encamped, was anew arranged, and Luckner marched upon Alfeldt, where he joined the Hereditary Prince's corps.

*Intended general attack.*

The Marechals now began to show symptoms of their speedy retreat. On the 13th the Saxons fell back to Lichtenau, on the 16th Munden and Gottingen were evacuated, and on the 17th placing ten battalions and three hundred cavalry in Cassel, their
whole

whole army marched to the left, and retiring between the Fulda and the Werra, reached the neighbourhood of Francfort about the 28th. Prince Ferdinand leaving a corps to watch Caſſel, accompanied their movements by Homburg, Schwartzenborn, Grebenau, Mahr, Ulrichſtein, Schotten, and Nidda, where he arrived on the 29th, and on the 30th was at hand to prevent the conſequences of our repulſe at Johanneſberg.

*Retreat of the French army.*

For during theſe movements of the principal armies, the Hereditary Prince having been conſiderably reinforced on the 20th, had advanced on the 21ſt to the heights of Homburg, diſlodged Levi on the 22d, and taken his camp and 500 priſoners. On the 23d he advanced to Grunberg to attack Condé, who fell back to Gruningen. —On the 25th he followed the enemy, whom he found too ſtrongly poſted to attempt with any proſpect of ſucceſs.—On the 26th as the Marechals were now approaching the Mayn, he retired upon Grunberg.—On the arrival of our army at Schotten on the 28th, he again advanced to Gruningen, and from thence on the 29th to Wolferſheim near Friedberg; beyond which and towards Rodheim Condé had retired to effect a junction with the two Marechals, who were now ſuppoſed to be at Bergen. On the 30th the Hereditary Prince ſent Luckner to poſſeſs the commanding heights of the JOHANNESBERG, while he ſhould march to Aſſenheim; but on receiving intelligence that Condé was advancing upon Johanneſberg, he altered his route to aſſiſt Luckner. The heads of the two corps met on the heights, where Condé (unexpectedly to us) being ſupported by the whole French army, ſoon obliged the Hereditary Prince, who was here ſeverely wounded, to retire upon Wolferſheim with the loſs of above two thouſand men, chiefly made priſoners.

*Movements of the Hereditary Prince.*

*Affair of JOHANNESBERG.*

This check, which encouraged the enemy for ſome time to act on the offenſive, ſuſpended whatever hopes Prince Ferdinand might have entertained of forcing them over the Mayn, or of gaining poſſeſſion of Friedberg, and taking a poſition which would have covered his future operations againſt Caſſel, Marburg, and Gieſſen. On the 30th he detached a conſiderable corps to ſupport the Hereditary

Prince, and on the 31st he himself marched forward to Bingenheim and Staden, in face of the French army, which after leaving the Saxons at Bergen was now encamped on the Johannesberg; and behind Friedberg.—On the 3d of September still desirous of preserving the appearance of superiority, he advanced Lord Granby's corps considerably on his left towards Bergen; but the enemy sensible of the advantage they had lately gained, began on the 4th to extend to their left, and on the 6th their whole army was in march upon Guningen—This determined Prince Ferdinand immediately to retire behind the Ohm, in order the more effectually to cover, and carry on the siege of Cassel, which was now his great object. He marched therefore on the 8th towards Grunberg, intending to have passed the Ohm the next day, but such was the extreme badness of the roads and weather, that it was the 11th before this could be accomplished by the artillery, and rear guard, on which the enemy pressed hard.

*Movements of the armies.*

In this manner did the first part of the campaign pass, in which Prince Ferdinand by a conduct singularly skilful had defeated a very superior enemy, overturned their schemes of conquest; and confining them in a narrow extent of ground behind a river, cramped them in all their future operations, cut off their communications, obliged them to call in another army to their assistance, and when that could not extricate them, by a series of masterly movements, severe and fatiguing to the troops, had forced them to abandon Hanover and Lower Hesse, and to retire upon the Mayn; this river possibly, they also must have crossed, but for the check received at Johannesberg, which induced him to fall back upon the Ohm, and to attend solely to the siege of Cassel.

*Advantages gained by Prince Ferdinand.*

For several days the enemy's motions behind Marburg, and to their left seemed to indicate an intention of forcing a passage towards Cassel by Frankenberg, and Corbach; but this at all events Prince Ferdinand was determined to oppose: therefore leaving his several posts on the Ohm of Homburg, Sweinsburg, Amœneburg, Kirchain, properly guarded, he extended to his right in several corps, and on the 15th obliged the French to quit the heights of Wetter and retire over the Lahn.—On the 18th, the enemy having marched to their

their right and encamped behind Amœneburg seemed thereabouts to threaten the passage of the Ohm, and occasioned the return of a part of our army towards that quarter.—Under Amœneburg which they now held invested they took possession of the BRUCKER MUHL, and of a small work on their side of a bridge over the Ohm, while we occupied a redoubt on ground some what lower fifty or sixty paces from the bridge on our side—On the morning of the 21st which was very foggy, they began an attack both with cannon and musquetry on our redoubt, we returned it, and insensibly a very warm local action was engaged, which lasted till it was dark. Seventeen battalions successively relieved each other in our redoubt, and twenty pieces of heavy cannon on each side were at last employed, from the distance of three hundred to five hundred yards—Night put an end to this singular and bloody contest, which was begun in the darkness of the morning from accident, but persevered in from mutual distrust and apprehension of a design entertained by neither side to force the passage of the bridge and river, and in consequence of which both armies were drawn out—Our loss in killed and wounded was about seven hundred, that of the enemy from being more exposed, was nearly double.

<sub>Affair of the Brucker Muhl.</sub>

During the remainder of this month, the armies made no considerable changes of situation; several efforts of the enemy to interrupt our communications were defeated by our light troops, and our position behind the Ohm was strengthned and secured by many works. The season advancing, forage becoming scarce, and Prince Ferdinand now feeling himself enabled to begin the siege of Cassel, which had been blockaded from the 17th of August, sent off the heavy baggage and useless horses on the 2d of October to Hameln—On the 11th, he reinforced Prince Frederick, who with twenty-six battalions and ten squadrons, opened trenches on the 16th at night; and on the 1st of November, the garrison capitulating with the honours of war, thereby allowed Prince Frederick's corps to join the army.

<sub>Surrender of Cassel.</sub>

The French on the 7th, knowing that the PRELIMINARIES of PEACE were signed on the 3d, proposed a cessation of hostilities; but

**Conclusion of the war.**

but Prince Ferdinand having received no such intimation from our ambassador at Paris, could not consent, and ordered seven battalions under Major Gen. Huth, on the 19th, to open trenches before Ziegenhayn. On the 14th, the courier from England arrived; on the 15th, the cessation was signed at the BRUCKER-MUHL; on the 16th, the French marched to their winter quarters; and on the 19th, the allied army separated. The British corps marched to their cantonments on the frontiers of Holland, through which country they began to pass on the 23d of January 1763, in several divisions; the last of which embarked from Wilhelmstadt on the 23d of March, and the whole arrived safe in ENGLAND.

**General observations.**

In this manner ended a WAR, glorious and honourable for PRINCE FERDINAND, and for the troops whom he commanded. The whole body of ALLIES served with distinguished reputation, and each successive campaign added fresh lustre to the British arms. With an army much inferior in numbers, but whose confidence in their general was unbounded; he successfully withstood for six years the fullest exertions of the French monarchy, and gained many signal victories. At the beginning of the campaign, he sometimes was obliged to give way to the strength and ardour of his enemies; but before the end, he never failed to resume the offensive, and deprive them of their short lived advantage. The difficulties that attend the conduct of an allied army, vanished before his superior reputation; and the singular instance of one without party, jealousy or discontent, was seen under his command. His arrangements for the supply of his troops, were just and well executed; his activity was unremitting; his local knowledge and combinations wonderful; his firmness and presence of mind were often tried; and the quick remedies which he applied to unforeseen misfortunes, or critical situations, marked his ready decision which was always conspicuous. From his well weighed purposes, no common difficulties ever diverted him, for he could depend on their being surmounted by the zeal of his generals, and the bravery and attachment of his troops. The theatre of war was new, and his system of operations was each year different; when he advanced, it was with vigour and effect, when he retired, he kept his enemy in respect; his defensive positions were most skilfully chosen, and of such a nature, that he was ever in a situation to take advantage of the opportunities which the moment might present. He did not always oppose the enemy in front; but skilfully covering and securing his own communications, he placed himself on theirs in such a manner, that when the country seemed most exposed and inviting them to take possession; they durst not advance for fear of having their retreat or their supplies cut off. There was no position from which his superior local knowledge and perseverance when on the offensive, did not enable him to turn and dislodge the enemy. The constant changes that took place from the defensive to the offensive war, and from the open to the woody and mountainous country, gave a full scope and display to the singular talents of this great man, who so successfully directed the several corps of an army, acting in concert from Saxony to the frontiers of Holland; and whose movements were frequently combined across the whole of Germany, with those of the KING of PRUSSIA his great master and instructor. All the actions he atchieved, may be traced up to his own superior skill and conduct; and in the rank of generals, PRINCE FERDINAND must always be considered as one of the most distinguished.

---

For much information, and many of the circumstances contained in this relation, Col. Dundas must here express his acknowledgements to his friend Major Gen. Roy; and most truly regrets, that one so perfectly acquainted with the scene of operations, and possessed of the most ample materials, should never have given to the world a detailed history of that war, the whole of which he served, and is so well enabled to investigate.

The annexed map will give some idea of the low, and of the mountainous country, as also of the tracts where the principal operations of the armies were carried on. Such names alone as are material are marked on it, because it was apprehended that the insertion of a greater number, would only tend to confuse a scale so small, and a map so slight.

The

## The BRITISH CORPS that served in GERMANY.

| General officers that arrived with the corps. | Numbers. | Regiments. | Periods of Arrival. |
|---|---|---|---|
| Duke of MARLBOROUGH<br>Lt. G. Lord G. SACKVILLE<br>M. G. Marquis of GRANBY | 2300 | royal regiment horse guards<br>1st } 3d } dragoon guards<br>2d } 6th } 10th } dragoons | These regiments under the command of the Duke of Marlborough, joined the army at the camp of *Coesfeldt*, in August 1758. |
| Waldegrave<br>Whiteford<br>Kingsley | 6200 | 12th<br>20th<br>23d<br>25th } infantry<br>37th<br>51st | |
| Mostyn | 1000 | Kieth's Highlanders | Joined in 1759. |
| Major Gen. Howard<br>Honeywood | 1300 | 2d dragoon guards<br>1st } 7th } dragoons<br>11th } | These regiments joined the army at the camp of *Wavern*, in May and June, 1760. |
| Douglas | 6200 | 5th<br>8th<br>11th<br>24th } infantry<br>33d<br>50th | |
| | 1000 | Campbell's Highlanders | |
| Eliott | 700 | 3d } 4th } horse | These regiments joined the army at *Saxenhausen* camp, in June, 1760. |
| Earl of Pembroke | 700 | 15th dragoons | |
| Cæsar | 2000 | 1st } 1st } battalions of guards<br>1st } | The guards joined the army at *Buhne* camp, in September, 1760. |
| | 500 | artillery | The artillery joined, and was increased from time to time. |
| | 22000 | | |

| British corps | 30 squadrons of cavalry<br>21 battalions of infantry and artillery | — | 5000<br>17000 | The grenadiers of the guards and line formed three battalions. |
| | | | 22000 | |

|  | Years | 1757 | 1758 | 1759 | 1760 | 1761 | 1762 |
|---|---|---|---|---|---|---|---|
| General establishment of the ALLIED army in Germany | Cavalry | 8600 | 13,160 | 14,060 | 18,160 | 18,160 | 18,400 |
|  | Infantry | 41,500 | 47,640 | 54,740 | 72,000 | 72,640 | 73,200 |
|  | Total | 50,100 | 60,800 | 68,800 | 90,160 | 90,800 | 91,600 |
| General establishment of the FRENCH army in Germany | Cavalry | — | 18,000 | 22,000 | 27,000 | 29,000 | 26,000 |
|  | Infantry | — | 82,000 | 90,000 | 109,000 | 112,000 | 100,000 |
|  | Total | 100,000 | 100,000 | 112,000 | 136,000 | 141,000 | 126,000 |

### Allies.

During the campaigns of 1757, 58, and 59, the light troops were gradually increased to about three thousand. In the campaigns of 60, 61, and 62, they were carried up to eight thousand. In the years 60, and 61, every effort was made to augment the army: it remained in 62, at the same high establishment, but was not then well recruited. The companies of grenadiers from the year 59, were formed into battalions, and reckoned as such in the strength of the army. Ten squadrons of Prussian dragoons served from 57, till 1760, and then joined the King; five squadrons of Prussian hussars served the whole war.

### French.

During the campaigns of 1757, 58, and 59, the light troops were gradually increased to about five thousand. In the campaigns of 60, 61, and 62, they were carried up to eleven thousand. In the years 60, and 61, the armies of France were greatly augmented; in the year 62, their numbers were rather lessened. In general, they were at least one half stronger than the allies. The establishment of the French battalions, was lower than that of the allies; but the cavalry per squadron was much the same.

|  | Nations. | Cavalry. | Infantry. | Total. | The Line. Bat. | Sqd. |
|---|---|---|---|---|---|---|
| General proportions of the ALLIED army in the Year 1761, including light troops. | British - - - | 5000 | 17,000 | 22,000 | 20 | 30 |
|  | Hanoverian - | 7600 | 27,000 | 34,600 | 30 | 34 |
|  | Hessian - - | 3600 | 18,800 | 22,400 | 34 | 16 |
|  | Brunswick - | 1200 | 8300 | 9500 | 11 | 4 |
|  | Buckeburg - | 60 | 740 | 800 | 1 | — |
|  | Saxe Gotha - | — | 800 | 800 | 1 | — |
|  | Prussian - - | 700 | — | 700 | — | — |
|  | Total | 18,160 | 72,640 | 90,800 | 97 | 84 |
|  | The Light Troops may be reckoned - - - - |  |  |  | 10 | 25 |

There were also five battalions of Hanoverian Militia in Garrison.

# PLATES.

The firſt poſition is expreſſed, by — — — } Yellow—or light dotted lines.

The laſt poſition is expreſſed, by — — — } Red—or a ſtrong ſhaded line.

The intermediate poſitions, are — — — { Blue—or two lines without ſhade.<br>Buff—or one ſingle ſtrong line.<br>Green—or two lines with a light ſhade.

Many of the poſitions are alſo marked 1, 2, 3, &c. in the order in which they follow.

The dotted lines and arrows expreſs the direction of movements.

The plates to No. 19, incluſive, ſhould be bound at the end of the work.

The plates from No. 20, incluſive, ſhould be bound at the end of the Appendix.

# References to Plate Nineteen.

## BRITISH.

|  | |  |
|---|---|---|
| *June* 19th to 21st. | | Camp of Newhaus, near Paderborn. |
| 21st to 22d. | A. | Camp of Stadt Geseke. |
| 22d to 23d. | B. | Bivouac at Alt Geseke. |
| 24th to 28th. | C. | Camp at Soest. |
| —— 28th. | D. | Camp of Werle, including the Hereditary Prince and Lord Granby's corps. |
| 29th to 30th. | E. | Camp at Lunderen. |
| *July.* 2d to 4th. | F. | Camp at Dortmundt. |
| 4th to 7th. | G. | Camp of Hemmerden. |
| 7th to 27th. | L. | Wangenheim's camp at Lipstadt. Camps of Hillbeck and Hohenover. Sporken's Camp at Hertzfeld. |
| —— 16th. | L. | General position of the army the 15th at night, and during the action of the 16th at Fellinghusen. |
| —— to 27th. | L. | Camp of Hohenover. |
| —— 27th. | M. | Camp at Borglen. |
| —— 28th. | N. | Camp at Erwite. |
| —— 29th. | O. | Camp at Stormede, from which the army marched to Buren next day. |

## FRENCH.

| |  |  |
|---|---|---|
| H. | Soubise's camp at Unna. | *June* 24 to *July* 4. |
| S. | Position of Soubise's army behind the Landwehr, light troops between Werle and Sheidingen. | 4th to 7th. |
| K. | Soubise's camp at Soest. Broglie's camp at Erwite. | 7th to 13th. 7th to 15th. |
| P. | Soubise's camp at Paradys. Broglie's camp at Ostinghusen. | 13th to 19th. 16th to 19th. |
| R. | General position of Soubise and Broglie's army, during the action of the 16th. | 16th. |
| K. | Soubise's camp at Soest to the 25th, when he retired across the Roer. | 19th to 25th. |
| | Broglie's camp at Erwite to the 26th, when he fell back upon Paderborn. | 19th to 26th. |
| | The enemy retired on the Dimel, and the Lower Rhine. | 27th. 28th. 29th. |

# PRINCIPLES

## OF

# MILITARY MOVEMENTS.

---

## APPENDIX.

WORDS of COMMAND, and Circumstances of EXECUTION, necessary in the Movements of the *Company, Battalion, Line, Open Column, Close Column, Echellon,* &c.

---

THE PRINCIPLES and OPERATIONS of the column, open and close, of the battalion, and of the line, have been detailed at length in the first part of this work; but in order to bring their several most material movements into more distinct view, it is here meant to recapitulate the words of command, and to subjoin the chief points of execution.

The following practical ABSTRACT is therefore given. *Division* is often meant as a general word for platoon, company, or whatever front the column then has. Columns are supposed marched off from the right; but the consequent changes of command and circumstances, that would attend columns marched off from the left, are sufficiently obvious. The words of command, which are twice marked, are first given by the leaders of battalions, and then repeated by those of divisions. The general heads of many changes of position are also added: their division and detail of execution, as well as those of other various combinations of manœuvre that may be readily imagined, are easily inferred from the EXAMPLES given.

The CIRCUMSTANCES which follow, are among many others, essential to be observed in movements and operations of the *column*, open and close, of the *battalion*, and of the *line*.

---

**Attentions in movement.**

ALL movements of the line, are in general regulated by a battalion of that flank which is nearest to, and is to preserve the appui. All movements of the column are directed by its head; and the commander of the whole is with the regulating body.

Every movement should be subdivided into its distinct parts, and each executed by its separate command.

All marches, movements, and formations, must be made on fixed points; and those points and lines must be previously and truly ascertained. Independent of distant objects, the given lines of march and formation must always be subdivided by mounted officers.

The necessity of field officers and adjutants, being at all times mounted and alert on horseback, is evident and indispensable; each battalion to manœuvre and act well in line, must have two mounted officers at least.

All alterations of position should begin from a previous halt, except giving a new direction to the heads of columns, or diminishing or increasing their front, which may be done while in motion.

**Commands.**

Words of command must be quick, loud, and in the instant circulated.

All alterations in carrying arms, change of pace, facing, inclining, halting, marching, and in general every operation of the battalion, whether in line or column, which ought to be executed by the whole at the same instant, are made in consequence of one word from the commander of each battalion: but when broken and in column, the leaders of divisions on many occasions, repeat or give the words of *march, wheel, halt*, &c. to their several divisions as is necessary.

**Music.**

Music and drums should never be used with a view to instruct, or to regulate the cadence or step of any body, great or small.

**Signals.**

Signals should not be multiplied, they are cautions addressed to the leaders only of divisions or battalions, who give their orders in consequence for the execution.

The ordinary march may be eighty in a minute, each step thirty inches in length: it is the pace on
all

all occasions, where greater celerity is not ordered. The quick march may be one hundred and twenty in a minute; each step thirty inches, when moving in front, and twenty-four inches when moving in file. It is the pace in all wheelings and filings of divisions from line to column, and from column to line; should it be increased beyond one hundred and fifty in a minute, it becomes a run. To these steps, and these only, and to their cadence, must the soldier be habituated without drum or music.

March.

In line, individual battalions must frequently by order, and without change of cadence, lengthen and shorten their steps to preserve dressing; and also on particular occasions, but for no great distance, larger bodies will be required to step out.

A division or company, may occasionally run; a battalion may march quick; but the hurrying a large body in front or in column, will certainly produce confusion and disorder: this is never to be risked where an enemy is to be encountered, though it may sometimes be necessary where a post or situation is to be seized.

Halting.

All *halts* are made to the point to which the troops are looking, when the order is given, and a separate word directs the after dressing.

When the word *dress* is singly given, it implies to the hand to which the troops are then looking. When eyes are to be turned to another point, the addition of *right*, *left*, or *center* will be used.

Dressing.

In dressing battalions, regard is had to the general line, not solely to a partial division of it.

No rank or body ought ever to be dressed without the officer on its flank, determining a given object on which the division or battalion is to be formed.

One flank at least of a line, battalion, or other body, is always considered as to be placed or posted near some support, which it ought not to quit. From that flank of *appui* the dressing always begins, and is made upon a point beyond the other flank in the direction in which the line is to be preserved. In general dressing, eyes are therefore always turned to the point of *appui*, which remains unmoveable, and from whence orders and directions proceed.

Wheeling.

All wheels or filings made from the halt into column or line, are made at a quick step.

In all wheels of the columns when in march, the wheeling flanks must proportionally quicken their pace, to prevent a stop in the succeeding

ceeding division; but this increased pace does not become necessary, if the wheel which the head makes exceeds not one-sixth of the circle.

During the wheel, eyes are turned to the wheeling hand, and after the wheel to the pivot flank.

A perfect uniformity in the formation and arrangement of all companies and battalions, as already prescribed, is indispensable for the execution of just and combined movements.

The equal strength of divisions, is essential to the correct manoeuvres of a battalion, or of a line.

**Column.** All changes of position from one point to another, are made in one or more columns.

Every column of march or manoeuvre is formed by a regular succession of the divisions, from right to left, or from left to right of the line, or of such of its parts as compose the column.

In column, divisions cover and dress to the proper pivot flank—to the left, when the right is in front; and to the right, when the left is in front.

**Open column.** The commanding officers of divisions (shift if necessary) to lead files, and also conduct pivot flanks, when marching in column in the alignement.

The countermarch of divisions is always to be made from the right, and behind the rear.

Columns of march or manoeuvre will generally be composed of platoons or companies—Close columns of companies, or grand divisions.

Columns of march or manoeuvre, will be formed with the left in front, whenever it is probable that the formation of the line will be required to the right flank; and vice versa when required to the left flank.

In column, rear ranks are one pace asunder—When a considerable distance is to be marched, they may be opened to two paces, but without increasing the distances of divisions, which remain such as are prescribed, according to the object of the movement. **Open column.**

All marches are made in columns of divisions, never by files. Filings are only applied to the internal manoeuvres of the divisions of the battalion, not to any considerable movements of the battalion itself, or of greater bodies.

In marching in column, the rear divisions follow every turn and twist, which the head one makes, until they are particularly ordered to gain a straight alignement; and that the object is to form in line.

In open column, the artillery, music, drummers, &c. wheel with, and remain closed up to the rear of their respective divisions. In column,

lumn, at half or quarter diftance, or in clofe column, they may occafionally file on the flank, which is not the pivot one. In the fquare or oblong, they will be in the interior part of it, or follow it when marching.

The column marching, at half or quarter intervals, fhould preferve a diftance betwixt battalions, equal to the front of the column.

In column of march, the leaders of divifions may be in front of their divifions, and the flank non-commiffioned officers preferve diftances; but when moving in an alignement, the leaders are then on the flanks, and become the pivots, who cover each other exactly—When the column halts, and the line is to form; the officers, and the fergeant coverers, place themfelves behind the files of men next to them, and thofe files (being then eafily corrected if neceffary, by the mounted officer) become the pivots on which the divifions wheel, and the line forms.

When marching in the alignement, there muft never be more than one file of officers on the pivot flank; all others are either on the oppofite flank, or in rear of the divifions: the colours cover the pivot files of men.

In whatever manner the leading flank of a battalion arrives in a determined line, a mounted officer always gives the precife point where it enters; and if it there halts, and is to form; he from thence if neceffary, corrects the pivot files of men on the given diftant points, before the divifions wheel up into line.

If the battalion after wheeling up from column into line, is not critically well dreffed, the fault muft be in the internal parts of the divifions; this muft be immediately corrected by each officer on the pivot men, who on no account muft move or fhift, but remain fo many given and fixed points, on whom the battalion is exactly lined.

When feveral columns of battalions are marching to form in one general line, a mounted officer fhould conduct the leading flank of each column—And diftances and dreffing of their heads, are taken from the regulating battalion, and the one adjoining to it.

All doubling up or increafing the front of the column muft be made before entering on the alignement.

All marching in the alignement muft be made in ordinary time, and taken up from the point where it is entered; the pivot officers then become more particularly anfwerable for the diftances and exact covering of the flanks.

If the line is to be formed to the flank, wheeling diftances muft be exactly preferved: if forming the line to the front is the immediate object, the column may march at half, or quarter diftance, and form either by deploying from clofe column,

*Open column.*

*Open column.*

lumn, or by wheeling into the alignement, taking diſtances, and forming up to the flank.

**Open column.**

When the column ſtanding with the right in front, wheels up and forms in line, the dreſſing of the whole will be of courſe to the right along the ſtanding pivots, and when the left is in front, the dreſſing will be to the left.

The point where the head of the column enters the alignement, which is given and muſt never be quitted by a mounted officer, but as he is relieved, and till the whole have entered it; the point where the flank of the leading battalion begins to form; the ſeveral adjutants who place themſelves in the true line; the colours of battalions which have halted, or are formed up; are ſo many marked points within the line itſelf, independent of diſtant objects, on which the dreſſing of pivots or battalions can be regulated.

The juſtneſs of wheeling diſtances and of leading upon the points of march; the covering of pivot flanks; and the true wheels of the quarter circle, are the indiſpenſable circumſtances that enſue exactneſs in movements of the open column, and in all changes and formations of the line.

When the cloſe column is halted, rear ranks are one foot, diviſions are one pace, and each battalion of which it is compoſed is three paces aſunder.

Muſic, drummers, pioneers, &c. are cloſed up to their ſeveral diviſions, or may be ordered on the flanks. Officers, ſergeants, &c. are in their places as in open column: ſtaff and artillery are on the flanks of the column.

The cloſe column ſhould not exceed five or ſix battalions; when there are more troops, more columns ſhould be formed.

**Cloſe column.**

The head of the cloſe column is always brought up, and halted in the line into which the column is to extend; and it muſt alſo be placed perpendicular to that direction.— Though the cloſe column may be required to march to the flank, yet a conſiderable movement in front cannot be expected from it without looſening the diviſions.

In echellon, the diviſions are retired at equal but parallel diſtances behind, but out-flanking each other.

**Echellon.**

Whatever diſtances the diviſions are retired behind each other; if thoſe diſtances are equal, and the diviſions alſo equal, the flank files of the whole will always line in a diagonal direction.

When the object is to form in oblique line; in proportion as the diviſions are retired, they muſt cover inwards, part of the one preceding them, viz. If retired a diſtance equal

to

to ¼—½—¾—of their front, they must cover respectively $\frac{1}{25}$—$\frac{1}{7}$—$\frac{1}{3}$, &c. of the front of the one preceding. In these situations, when the divisions wheel, the line will be formed without false intervals, and more or less oblique, as required.

The whole or only part of the body may be thrown into echellon, and that either to the front or rear. In the one case to gain a flank, in the other to refuse it.

The echellon may be formed on a flank, or on any central division, either marching or halted, to front or rear.

The echellon positions and movements are necessary and applicable to a considerable corps, rather than to a battalion, which may however occasionally assume that shape for the sake of instruction.

The exactness of march in line, depends on the squareness of each individual's body, on the touch of the files, on the uniform cadence of step, and on the exact perpendicular of march, given by the advanced colour and sergeants, which the battalion in every respect covers, follows, and complies with.

The cadence is not to be altered by particular battalions when marching in line, but when it is necessary, they may lengthen or shorten their step, by word from their own commanders.

The whole line should halt at the same instant when the word is given, and no dressing or correction of intervals takes place till so directed—The advanced colour on the halt falls back to the battalion.

The line retires by the advanced colours, in the same manner that it advances.

Distances of battalions are taken from colours to colours, and the mounted officer in the rear, can best direct their preservation.

The march and halt, and attention of each battalion in line, is by its own center. The commander alone regards the regulating battalion—Dressing to a flank is by a separate direction, and given when necessary or proper after halting.

The intervals betwixt battalions are twelve paces—When without cannon, they may be six paces.

Cautions and words of command, are repeated loud by commandants of battalions, from the one regulating—the words, *march*, *halt*, &c. of each, must be instantaneously circulated.

When the line is marching, there may be two regulating colours of two adjoining battalions, with which the others are apprized always to dress—The commander remains on the flank of one of them, and from thence by a signal of his hand, directs the other to lengthen or shorten the step, so as to preserve the parallel movements of the line:

*March in line.*

The march of a confiderable body in line, can only be at the ordinary ftep; a quicker movement would produce diforder; nor could artillery attend its motions when advancing to the enemy—But there are fituations where a brigade or fmaller front, fhould move on to a particular attack at a lengthy ftep, or where even a quicker cadence may be required from them.

## COMPANY.

Commands given by Leaders of

| Nature of Movements. | Battalions. | Companies or Divifions. | Circumftances of Execution. |
|---|---|---|---|
| When the company being halted is to march in front. | | *Forward* | A caution—The points of march are given to the leader. |
| | | *March* | Dreffing to the right, or by whatever point is directed—The flank officer conducting on the fixed objects. |
| When to halt. | | *Company Halt* | A caution To which ever hand it is then looking while in march. |
| | | *Right, Drefs* | If neceffary. |
| When halted and ordered to drefs. | | *Drefs* | Eyes are turned to the hand ordered—Each individual dreffes, fo as juft to difcover the face of the 2d man from himfelf; and the flank officer corrects the whole of the front rank, on the given diftant object. |

When

| Nature of Movements. | Commands given by Leaders of | | Circumstances of Execution. |
|---|---|---|---|
| | Battalions. | Companies or Divisions. | |
| When halted, and to wheel by platoons to either flank into column. | | *By Platoons to the right, wheel.* | A caution—The pivot man of the front rank of each platoon, faces square to the right—A non-commissioned officer of the right advances, and marks the ground at which the left of his platoon is to arrive. |
| | | *March* | Eyes to the left, platoons wheel quick, and with an uniform front. |
| | | *Halt* | Halt to the left when the quarter circle is compleated, officers are placed in the front of their platoons, or on the pivot flanks. |
| When platoons are halted in column and wheel up to form in line. | | *Platoons to the left wheel to form in line* | A caution—Pivot man of each platoon faces square to the left. |
| | | *March* | Eyes to the right. |
| | | *Halt* | Each platoon separately by the right and officers take their posts on the right. |
| When to march in file. | | *To the right, face* | The whole face to the right. |
| | | *March* | At a short quick step, in file, opening out as little as possible, and exactly covering the leader who conducts the body truly on two distant points. |

( 12 )

Commands given by Leaders of

| Nature of Movements. | Battalion. | Companies or Divisions. | Circumstances of Execution. |
|---|---|---|---|
| When filing, and ordered to halt and front. | | Halt | The whole file covering exactly, and at just distances; the leader dressing it exactly on the intended point before fronting. |
| | | Front | By facing to the left and dressing to the right. |
| When to countermarch. | | *The company will countermarch* | A caution—A non-commissioned officer from the right, shifts to where the right is to be placed. |
| | | *To the right, face* | The whole face to the right. |
| | | *March* | In file behind the rear rank, till the right flank arrives at its ground where the left stood. |
| | | Halt | The leader then halts it, and makes the files dress and cover exactly in the new direction. |
| | | Front | The whole face to the left, and dress to the right. |
| When the comany is moving and is or- | | *To the right incline* | At this word, the whole will look to the right, preserve the body perfectly square to the proper front, and step with the right foot to the right, in the direction of the incline, and the left foot brought exactly before it.<br>The company must not open or fall into file, and must move in a direction perfectly parallel to the one they have quitted—The leader must |

dered

Commands given by Leaders of

| Nature of Movements. | Battalions. | Companies or Divisions. | Circumstances of Execution. |
|---|---|---|---|
| dered to incline. | | | be very exact in taking his oblique points of marching, and not exceed an angle of about forty degrees with the former perpendicular of march. |
| | | *Front* *Forward* | The whole cease inclining and move forward in line, parallel to their first direction. |

## BATTALION or LINE.

Commands given by Leaders of

| Nature of Movements. | Battalions, | Divisions. | Circumstances of Execution. |
|---|---|---|---|
| When the battalion is halted and is to march in front. | *The battalion will move forward* | | A caution—The points of march are given as directed, after the 1st colour and the sergeants covering them have moved out three paces. The colour is betwixt the sergeants, its place in the front rank is supplied by the second colour moving up from the center rank. |
| | | *March* | The whole looking to the center, which follows the advanced colour at the distance of three paces, and is directed by the commanding officer in the front, and major in the rear. |

When

( 14 )

| Nature of Movements. | Commands given by Leaders of ||  Circumstances of Execution. |
|---|---|---|---|
| | Battalions. | Divisions. | |
| When the battalion is to halt. | *Battalion* | | At this word the colours cease to advance; the battalion marks the time, and dresses well by the second colour. |
| | *Halt* | | The whole halt, looking to the center—Colours and sergeants resume their places. |
| | *Dress* | | The center, which cannot be wrong, does not move—And the dressing is obtained as directed for the battalion. |
| | *Eyes to the right.* | | The whole look to the right. |
| | | | When several battalions march in front, the advanced colours of the whole are kept as much as possible in proper line, by direction from the leader of the flank regulating one. |
| | | | When they halt, he will then be particularly careful, that the adjoining colours stop in the true intended line, and he will dress the line in the manner prescribed under that article. |
| When the battalion is halted, and is to march in file. | *To the right face.* | | The whole face to the right: the officers move out three paces to the left flank, and the covering sergeants take their places in the ranks. |
| | *March.* | | The whole step off at a quick short step, each man replacing the foot of the one before him, and the leader conducting on two distant objects. |

The

( 15 )

| Nature of Movements. | Commands given by Leaders of | | Circumstances of Execution. |
|---|---|---|---|
| | Battalions. | Divisions. | |
| | *Halt.* | | The battalion must not open out; nor ought it ever to march any considerable distance in this manner, and that chiefly for the purpose of opening or closing an interval in line. If more ground is to be gone over, it should be done by division marching. |
| | | | The whole covering truly, and at just distances. |
| | *Front.* | | By facing to the left, and dressing to the right.—The officers resume their places in the front rank. |
| When the battalion is moving, and is ordered to incline. Fig. 58. | *To the right incline.* | | As directed for the company. |
| | *Forward.* | | The whole look to the colours, and move forward in line ~~parallel to~~ their former front. |
| | | | If the intention of the incline is only to regulate an interval when moving in line—Heads will remain turned to the colours during such operation. |
| When the battalion is marching in front, and makes a small change in its direction to either flank. Fig. 60. | | | The colours by the most insensible alteration of the position of the persons of their bearers, take new points of march, and move upon them. One half of the battalion gives back, and the other gradually advances, till the whole are again perpendicular to the points of march. |

When

( 16 )

Commands given by Leaders of

| Nature of Movements. | Battalions. | Divisions. | Circumstances of Execution. |
|---|---|---|---|
| When the battalion is halted and is to retire. | *The battalion will retire* | | A caution. |
| | *Right about face* | | Dressing to the former point now the left. |
| | *Forward* | | Colours and sergeants move out, and points of march are taken as directed for the battalion. |
| | *March.* | | |
| When the battalion after retiring, halts and dresses. | *Battalion.* | | A caution—it ceases to advance and dresses. |
| | *Halt.* | | By the center. |
| | *Right about face* | | By the center, colours resume their places. |
| | *Dress.* | | As directed for the battalion. |
| | *Eyes to the right.* | | If necessary, and that the battalion is to remain halted. |
| When from battalion, the divisions retire by files from the right, and again front in column and wheel up into line. Fig. 126. | *The line will retire by files from the right of divisions.* | | A caution. |
| | *To the right face.* | | The whole, and the heads of files, disengage themselves to the right. |
| | *From the right of divisions, file.* | | A caution. |
| | *March* | | The whole march to the rear, regulated by the right division. |
| | | *Halt.* | Each division taking it from the head, and the rear of the files closing up and covering. |
| | | *Front.* | The whole are then in column. |

To

| Nature of Movements. | Commands given by Leaders of | | Circumstances of Execution. |
|---|---|---|---|
| | Battalions. | Divisions. | |
| To the left wheel and form in line | | Left dress | To the pivot flanks which are then dressed. |
| | | | A caution. |
| | | March | As before—In forming the line from column. |
| | | Halt | |
| | Right dress. | | |
| When the battalion is advancing in front and any partial obstacles occur to interrupt parts of it. Fig. 61. | | Company Halt | Such parts of the line as are not interrupted, still move on—Such platoons or companies as cannot continue their march in front, are ordered to halt. |
| | | To the right, or left face | The platoon will *file* from one flank—The company will file from both flanks—And two contiguous companies thus interrupted, will each file from its outward flank. In this manner, they follow the adjoining parts of the line which are marching in front, and which take care to preserve the intervals that the filing divisions should occupy. |
| | | March | |
| | | | If a half or whole battalion is thus interrupted, instead of filing, it may follow in two columns of subdivisions from its outward flanks. While the battalion is advancing, in proportion as the obstacle increases or diminishes, will the formed and filing parts of the line increase or diminish. |

P p

( 18 )

| Nature of Movements. | Commands given by Leaders of | | Circumstances of Execution. |
|---|---|---|---|
| | Battalions. | Divisions. | |
| When the obstacle is passed and the line is again completed | | *Move into line*<br><br>*Quick*<br><br>*Dress* | The intervals having been preserved; as soon as the obstacle in whole or in part is passed, the divisions which are in file or their parts successively, are ordered to move up quick, and resume their places in line. |
| When the line is retiring, and an obstacle is to be passed.<br><br>Fig. 61. | | | As when advancing in line, the interrupted parts follow in file, the formed parts; so when *retiring* (and when an enemy is near) they precede the line, which must therefore halt, when such operation is performing by any considerable part of it, to prevent too much hurry. In both cases, the natural order of the battalion is preserved, and the flank of the entire part must be joined by the proper flank of the broken part: the filing must therefore begin accordingly; in retreating from the point which is farthest removed from the part that remains formed; and vice versa when the line is advancing.<br><br>As soon as the ground admits, the broken parts re-enter the line; in the one case by waiting for it, and in the other by moving up quick.<br><br>But if in retiring, no particular attempts of the enemy are to be apprehended; then the breaking off may be performed behind the line, in the same manner, as when advancing; and the halt and inconveniences occasioned by the breaking off before the line, will be much avoided. |

The

( 19 )

| | |
|---|---|
| Passage of battalions or lines through each other. | The second line advances to within twenty yards of the first, and halts. The first line then receives the word *pass*—The whole face to the right, and each platoon disengages its head—They then march, pass quick through the second line (which throws back four files wherever the heads of platoons present themselves to pass) and when through, resume the ordinary step—The heads of files keep drest, and at a given distance of one hundred and fifty or two hundred paces, they halt, front into column, and wheel up into line.<br><br>Should the second line remain posted, the first retires till within twenty yards—At the word *pass*, each platoon turns to the left, passes from the left through the second line, and proceeds to form as above. |

---

## OPEN COLUMN.  The RIGHT in FRONT.

Commands given by Leaders of

| Nature of Movements. | Battalions. | Divisions. | Circumstances of Execution. |
|---|---|---|---|
| | *By divisions to the right wheel* | | A caution—The pivot man of the front rank of each division faces square to the right.—A non-commissioned officer of the right division runs out to mark the ground at which his left flank is to arrive. |

( 20 )

| Nature of Movements. | Commands given by Leaders of | | Circumstances of Execution. |
| --- | --- | --- | --- |
| | Battalions. | Divisions. | |
| When the battalion is halted, and to wheel by divisions of any kind into column, to either flank.<br><br>Fig. 13. | March | | Eyes to the left, step off quick, and wheel with an uniform front. |
| | | Halt | By the left when the wheel of the quarter circle is completed, each division separately. Officers place themselves in front of their divisions, if the continuation of the march is the object; or on their pivot flanks, if the column is to march and form in the alignement. |
| | | Dress | Eyes to the pivot flank. |
| When the battalion is to march in open column of divisions from a flank to the front.<br><br>Fig. 11. | | Flank division | A caution—Whether right or left. |
| | By divisions to the right wheel | March | The leading flank division only advances a space equal to its own front, and then halts. |
| | | Halt | |
| | March | Halt | The rest of the battalion wheels the quarter circle to the right by divisions, and each separately halts. |
| | March | March | The whole column is put in motion, and follows the direction which the front now takes; officers being in front of their divisions, or on their pivot flanks, according to the object of the march. |
| When the battalion is to march in open column of divisions from the flank to the rear.<br><br>Fig. 12. | | | The whole battalion is wheeled to the flank by divisions, and halted. |
| | | | The front division is then ordered to continue its wheel to the rear, and is followed by the rest of the column which is put in march. |

When

| Nature of Movements. | Commands given by Leaders of | | Circumstances of Execution. |
|---|---|---|---|
| | Battalions. | Divisions. | |
| When the battalion is to form open column to the front on any one division. Fig. 9. | *The battalion will form open column to the front on the 5th division—The right in front* | | A caution—To form at half, quarter, or whole distance. |
| | | *Face inwards to the 5th division* | The fifth division stands fast.—The rest of the battalion faces to that division, and the heads of divisions disengage themselves to the right. |
| | | *March* | The whole file quick—To their places in column. |
| | | *Halt* | The head of each division, when in line with the flank of the fifth.—The rear of the division closing in. |
| | | *Front* | Each division is then at its proper distance and place in column. |
| | | *Eyes to the left* | Left pivots cover, as the right is in front. |
| When the column is halted and is to march. | *The column will move forward* | | A caution.—Its points of march are ascertained to the leader. |
| | *March* | *March* | The word is repeated quick from front to rear of the column by the several leaders of divisions. |
| When the column is marching and is to halt. | *Halt* | *Halt* | Beginning at the front when arrived at its ground and repeated instantaneously by each leader to his division.—The pivot flanks must carefully cover in the true line when they halt. |

When

Commands given by Leaders of

| Nature of Movements. | Battalions. | Divisions. | Circumstances of Execution. |
|---|---|---|---|
| When divisions are halted in column, and wheel up to either hand to form in line. Fig. 14. | *Divisions to the left wheel into line* <br><br> *March* <br><br><br> *Dress* |  <br><br><br> *March* <br> *Halt* <br><br> *Dress* | A caution—Pivots being dressed, and the divisions at due distances, each flank man of the front rank faces square in the new line. <br> Eyes to the right. Step off quick. <br> Each division separately by the right, in the line of the pivot men. <br> Which can hardly be necessary, if the whole have lined with, and halted on the pivot men who remain immoveable.—Officers take their proper places, if not at them. |
| When the head of the column wheels into a new direction, marches on, and is successively followed by the rear divisions. Fig. 129. |  | *Leading Division.* <br> *Halt* <br> *Right wheel* <br> *March* <br><br> *Halt* <br> *Dress* <br> *March* <br><br> *Succeeding Division.* <br> *Halt* <br> *Right wheel* <br> *March* <br> *Halt* <br> *Dress* | When the leading division of a column arrives at the point A, where it is to wheel—It will receive the words *halt, wheel, march*; and its outward flank will then increase its pace, at least in the proportion of the circumference of the wheel, to the distance betwixt divisions. <br> When the wheel is completed, it will receive the words *halt, dress*, and then instantaneously the word *march*, to resume the general pace at which the column is marching. <br> In this manner the pivot point A will be free for the succeeding division, which comes up to it at the general pace, and there receives the words *halt, wheel, march, halt, dress*. Each makes a pause after the word halt, and steps off at the word march, given |

| Nature of Movements. | Commands given by Leaders of | | Circumstances of Execution. |
| --- | --- | --- | --- |
| | Battalions. | Divisions. | |
| | | March | given when at its due distance, from the preceding division, and the instant before the coming up of the succeeding one. |
| | | | At the word *wheel*, eyes will be always directed to the wheeling flank. At the last word *march*, heads will be turned to the pivot flank, as the division is then to move on. |
| Wheel of unequal divisions in column of march.<br><br>Fig. 130. | | | As the divisions in column always cover on the pivot flank, when they are equal in extent, each will wheel on the exact ground of the preceding one. But when they are unequal, the wheel of each will be calculated to begin, so that at its completion, the pivot flanks may still cover in the true line.<br>In this manner, if the column is marching with the right in front, and that the wheels are to be made to the right—The second division (which is supposed out-flanked four files by the first and third) will pass the ground on which the first wheeled, a space equal to the extent of four files before it begins to wheel at a—And the third division will begin to wheel at a like distance, short of where the second division wheeled, viz. On the ground of the first to which it is equal in extent—This difficulty does not take place in wheels to the pivot flanks, which continue to cover both in wheeling and marching.<br>When the column has its left in front, the pivots then dress to the right, and the out-flanking of divisions will be to the left. |

| Nature of Movements. | Commands given to Leaders of ||| Circumstances of Execution. |
|---|---|---|---|---|
| | Battalions. | Divisions. |||
| When the leading division of a battalion marching in column, has wheeled into a new direction, and *halts* for the purpose of forming in line; the other divisions then instantly *file* to cover it in column, and are in readiness to wheel into line.<br><br>Fig. 18. | Halt | Leading Divisions. | *Halt*<br>*Right wheel*<br>*March*<br>*Halt*<br>*Dress*<br>*March* | The leading division wheels into the new line—Marches with its left flank along it—and halts when it arrives at its intended point of formation C. It is followed successively by the other divisions of the column. |
| | | | *Halt* | The whole divisions of the battalion halt when the leading one halts. |
| | | Succeeding Divisions. | *To the left face*<br>*March* | But at the instant that the head and such other divisions as have wheeled into and are now in the new direction do halt; each other division which is still marching in the old column is ordered to face to the left, and is separately conducted in file to gain by the shortest route its point of formation in the line. |
| | | | *Halt* | Each division, when its left flank arrives in the new line, and at its proper distance from the one preceding it. |
| | | | *Front* | When the rear of each division has closed up—Dressing is to the left which is the pivot flank, and the whole column is now in a situation to form in line. |
| | | | | The battalion may now form in line by the divisions wheeling up to the left. |

| Nature of Movements. | Commands given by Leaders of | | Circumstances of Execution. |
|---|---|---|---|
| | Battalions. | Divisions. | |

| Nature of Movements. | Battalions. | Divisions. | Circumstances of Execution. |
|---|---|---|---|
| When in open column of several battalions marching by divisions, one or more having wheeled into a new direction, halt, in order to form in line—The others continue their march in separate columns, to gain such new direction and be in readiness to form line.<br><br>Fig. 26, 27. | | *Leading division of each battalion, to the left wheel*<br>*March Forward* | When the head of the column is arrived at its point of formation, the whole receive the word *halt*.<br>Such divisions of the last battalion which has begun to enter, but are not yet all in the alignement, are immediately ordered to march to the flank as before directed, and place themselves in column.<br>Such intire battalions as are still in the old line of march, are at the same instant directed into the alignement in the following manner:<br>They break separately from the general column to the left, and each directs its front division on its respective point in the new line, the rear divisions following in column the tract of their leading one. |
| | | *To the right wheel*<br>*Halt* | When the head of each battalion arrives behind the new line, it wheels to the right, with its left flank placed in it—halts—dresses—covers to the left. |
| | | *Halt* | The remainder of each battalion when its head halts. |
| | | *To the left face*<br>*March*<br>*Halt*<br>*Front* | When the leading division of each battalion thus halts in the new direction, the rear divisions then are ordered to gain it by *filing* to their flank, where they halt and front, with their left flanks covering in the line. |

( 26 )

| Nature of Movements. | Commands given by Leaders of | | Circumstances of Execution. |
| --- | --- | --- | --- |
| | Battalions. | Divisions. | |
| | | | As each battalion is halted in column in the new line, it may separately wheel up by divisions and form in line. |
| When several battalions halted in column at half or quarter distance, are to form in line to the front of their march, in any oblique position. Fig. 50. | | | The whole may close up to quarter distance at least—The battalion which is to form on the nearest point of the line will be named—The several battalions will disengage to right and left, from the general column, and stand in an echellon position—The head of each battalion will then be directed to its nearest flank point in the new line, will enter it, take its open distances, and form in line, by wheeling up of divisions. Or—each battalion when arrived at its flank point, will halt in close column, and deploy into line. |
| | *The divisions of the column will countermarch, each form its right behind its own rear.* | | A caution—For each division to countermarch separately. |

When

Commands given by Leaders of

| Nature of Movements. | Battalions. | Divisions. | Circumstances of Execution. |
|---|---|---|---|
| When the column is to change its front by the countermarch of divisions. Fig. 20. | *Right face* <br> *March* | *Right face* <br> *March* <br> *Halt* <br> *Front* <br> *Dress* | Each division—An under officer marks where the leading flank of each division is to arrive. <br> Each division files to its right behind its own rear. <br> Each division when it comes to its ground. <br> Front to the left, and dress to the right, which is now the pivot flank, as the column has its left in front. |
| When the column, with its head arrived at the new direction, is to form in line facing to its then rear, and on the leading division. Fig. 39. | | | The same operation takes place as before directed in forming the line to the front, except that the front divisions arrive with their left flank before the intended line, and wheel to the left. The rear divisions, and battalions, *file* and *march* to the right, when the head of the column is halted, in order to gain the new position. |
| When the column is halted and it is necessary to form the line on the rear division, facing either to the then front or rear of the column. Fig. 17. | | | In either case, the column will first countermarch by divisions singly, and become a column with the left in front.—The right being now the pivot flank. <br> In the first case, the leading division will wheel to the right; and in the second to the left, in order to march along the intended line. <br> In both cases the line will be formed by the wheeling of divisions to the right. |

When

| Nature of Movements. | Commands given by Leaders of | | Circumstances of Execution. |
|---|---|---|---|
| | Battalions. | Divisions. | |
| When the column is to form in line to the right flank. Fig. 128. | | | 1. If the column is halted, and adjusted, it may be done by wheeling of divisions to the right into line; but then both battalions and divisions of battalions will be inverted. Or, 2. If the line is first formed to the left, and then each battalion countermarched, it will front to the right, and prevent the inversion of divisions in the battalion; but not that of battalions in the line. This can only be remedied by 3. First forming to the left, and then countermarching the line. Or, 4. The whole column must first countermarch by divisions in front to the rear, and then wheel to the left into line. |
| When a battalion in column of divisions files to the right to change position, and again fronts to resume column. Fig. 30. | | *Right face* | The whole divisions of the column being then in file. |
| | | *March* | The whole to the right flank, being regulated by the two leading divisions which march steadily on their given points A. C. |
| | | *Halt* | When the leading division arrives at its point in the new line it halts.— The others also successively halt, when at their ground.—The rear of the files close up, and their heads are dressed. |

*Front*

| Nature of Movements. | Commands given by Leaders of | | Circumstances of Execution. |
|---|---|---|---|
| | Battalions. | Divisions. | |
| | | *Front* | When the leading division fronts, the others separately front. Pivot flanks are dressed, distances are just, and the column is again in a situation to form in line if ordered.<br><br>The column would file to the left if it was to change position to that flank. |
| When a column of several battalions marching by divisions, changes position to a flank.<br><br>Fig. 30. | | | The preceding regards a column of one battalion only. The same operation may however take place in some situations when the column is composed of a greater number. But in general in such case the front battalion will alone *file* as mentioned, and the other battalions will each break from the original column, and march separately in *column* of divisions to their point in the new line. When the leading division of each wheels into it, the rear divisions of each then file, and take their places in the new column, ready to wheel up into line. |
| When the column of one or more battalions is to form in line on a central division in any given direction.<br><br>Fig. 131. 132. | | | The named division is placed with its pivot flank perpendicular to the intended direction. All the divisions in front of the named one countermarch and face it. The divisions of the named battalion file into the new column, the other battalions march separately in columns of divisions into it. The whole place their pivot flanks in the new direction, and form in line by wheeling up. |

When

|   | Commands given by Leaders of | | |
|---|---|---|---|
| Nature of Movements. | Battalions. | Divisions. | Circumstances of Execution. |

| Nature of Movements. | Battalions. | Divisions. | Circumstances of Execution. |
|---|---|---|---|
| When the line changes position on any central division.<br>Fig. 127. | | | The named division is placed with its pivot flank perpendicular to the new direction. The whole line wheels inwards by divisions towards the named one, and proceeds as already directed for the column. |
| When the column forms in line on its front, or rear division.<br>Fig. 133. 134. | | | The front division is placed with its pivot in the new direction; the remainder of the column moves to the flank and covers it. The line is formed by wheeling to the pivot hand.<br>If on the rear division, all the others must first singly countermarch, and then proceed as before. |
| Fig. 131. A. | *The line will form oblique to the left of the rear, on the 5th division of the 2d battalion.*<br>*The divisions of the line in front of the named division countermarch.* | *Right face*<br>*March*<br>*Halt*<br>*Front* | A caution.—The named division is placed perpendicular to, and with its left flank in, and behind the new line of direction, and fronting to the right of the new line.<br>Viz.—The first battalion, and the four right divisions of the second, in order to face the named division.<br>The several divisions of the right of the line, who will, when countermarched, face the named division in column. |

To

( 31 )

| Nature of Movements. | Commands given by Leaders of | | Circumstances of Execution. |
|---|---|---|---|
| | Battalions. | Divisions. | |
| To form an oblique line, facing to the left of the rear on a central division. The front of the column thrown to the left.<br><br>Fig. 131. A. | Central battalion. | Right face | The leaders of divisions of the second or central battalion only give this word. All the divisions of that battalion, except the named one, face to the right, in order to point to their places in the new column. |
| | The line | March | Given by leaders of divisions of the whole line.—Those of the central battalion file.<br><br>The leading division of every other battalion wheels, marches in front, and is directed towards its point in the new line, and is followed in column by the other divisions of that battalion. |
| | Central battalion | Halt | Each division of the central battalion halts in file, when the right flanks of its right divisions, and the left flanks of its left divisions are arrived and placed at their due distance in the new direction.—Such flanks being regulated by the commanding officer who is with the named division. |
| | | Front | Viz.—The divisions of the central battalion, the left of which are in the rear of the 5th division, and the right of which are in its front—These last face the 5th division, and therefore the division next it has preserved a double distance in taking its place in the new column, that such interval may be filled up by the wheel into line. |

The

| Nature of Movements. | Commands given by Leaders of | | Circumstances of Execution. |
|---|---|---|---|
| | Battalions. | Divisions. | |
| Fig. 131. A. | *The central battalion will form in line.* | | A caution to the battalion. |
| | | *Wheel inwards* | A caution to the divisions of the central battalion. |
| | | *March* *Halt* | And form in line. |
| | | | The central battalion being thus formed, the other battalions will proceed to form as follows. |
| | | *Front division to the right* *Wheel* *Halt* | By the time the leaders of the central battalion divisions have taken their places in the new column, the heads of the other battalions will be arriving in the new line, and each will wheel into it, and receive the word halt. The right battalion so as to have its right flank in the new direction—The left battalions so as to have their left flanks in it. |
| | | *Halt* | The rear divisions of each battalion. |
| | | *Rear divisions to the right face* *March* *Halt* *Front* | When the front division of each battalion halts in the new direction, all the following divisions are ordered to face, file, and each halts when its pivot flank is in the new direction, and fronts by facing to the left. |
| | | | Thus battalion after battalion arrives and takes its place in the new column. |

( 33 )

| Nature of Movements. | Commands given by Leaders of ||  Circumstances of Execution. |
|---|---|---|---|
| | Battalions. | Divisions. | |
| Fig 131. A. | Wheel to the { Right and / Left } form line | Right / Left } wheel<br>March<br>Halt | From new column, the whole or parts of it are ordered to form the line. The divisions of the central battalion as above, wheel inwards, both to the right and left; those of the right battalion wheel to the right, those of the left battalion wheel to the left.—The several explanatory commands of right and left are given according to the position of such battalions. |
| | Right dress | | If ordered to the right, or any other particular point, as the line will then be halted and looking to the named division. |
| To form an oblique line on a central division facing to the left of the front, the head of the column being thrown to the right.<br><br>Fig. 131. B. | The line will form oblique to the left of the front on the 5th division of the second battalion.<br>The divisions of the line to the right of the 5th division will countermarch. | | The named division will be placed with its left flank in, and behind the new direction, and fronting the right of the new line.<br><br>A caution. |
| | | Right face<br>March<br>Halt<br>Front | Those divisions.<br><br>Each other circumstance and word of command is exactly the same as when the oblique line is formed, facing to the right of the front. |

R r

( 34 )

Commands given by Leaders of

| Nature of Movements. | Battalions. | Divisions. | Circumstances of Execution. |
|---|---|---|---|
| To form an oblique line on a central division facing to the right of the front—The front of the column being thrown to the right of the rear. Fig. 132. **G**. | The line will form obliquely to the right of the front on the 5th division of the second battalion. | | The named division will be placed as before, its left flank in and behind the new direction, and fronting the right of the new line. |
| | The divisions of the line to the right of the 5th will countermarch. | | A caution. |
| | | Right face March Halt Front | Those divisions. |
| | Central battalion | Left face | The divisions of the central battalion only. |
| | The line | March | The whole—The central battalion files; the others wheel and march in front, in column of divisions separately. |
| | Central battalion | Halt Front | The divisions of the central battalion only. The right divisions being then in the rear, and the left divisions in front of the named one. |
| | The central battalion will form in line | | A caution to the battalion. |
| | Wheel inwards | | A caution to the divisions of the central battalion. |
| | | March Halt | And form in line. |

To

| Nature of Movements. | Commands given by Leaders of | | Circumstances of Execution. |
|---|---|---|---|
| | Battalions. | Divisions. | |
| To form an oblique line on a central division facing to the right of the front —The front of the column being thrown to the right of the rear. Fig. 132. C. | | | The central battalion being thus formed, the others will proceed to form as follows. |
| | | *Front division to the right wheel Halt* | Of each battalion when it arrives in the new column. |
| | | *Halt* | The rear divisions of each battalion. |
| | | *Rear divisions to the left face March Halt Front* | The rear divisions of each battalion, when the front is arrived at its ground. |
| | Wheel to the {Right / Left} and form line | | A caution to each battalion. |
| | | *Right / Left wheel* | A caution to the several divisions. |
| | | *March Halt* | The divisions of the several battalions. |
| | *Right dress* | | The several battalions. |

| Nature of Movements. | Circumstances of Execution. |
|---|---|
| To form an oblique line on a central division, facing to the right of the rear—The front of the column being thrown to the left of the rear.<br><br>Fig. 132. D. | The named division will be placed with its left flank behind the new direction, and fronting the right of the new line.<br>The divisions in front of it will countermarch and arrive with their right flanks, as those in rear of it will arrive with their left flanks behind the new direction.<br>The line will be formed by the divisions wheeling to the intended front. |
| To form an oblique line on the front division, facing to the right of the front—The rear of the column thrown to the left of the front.<br><br>Fig. 133. E. | The front division will be placed with its left flank behind the new direction.<br>The rest of the column will march to the left, and place the left flanks of their divisions behind the new direction<br>The line will be formed by divisions wheeling to the left. |
| To form an oblique line on the front division, facing to the left of the front—And the rear of the column thrown to the left.<br><br>Fig. 133. F. | The same operation as in facing to the right of the front. |
| To form an oblique line on the front division, facing to the left of the rear—The right of the column thrown to the right.<br><br>Fig. 133. G. | The front division will be placed with its left flank behind the new direction—The rest of the column will march to the right, and place the left flanks of their divisions behind the new direction—The line will be formed by wheeling to the left. |

| Nature of Movements. | Circumstances of Execution. |
|---|---|
| To form an oblique line on the front division, facing to the right of the rear—The rear of the column thrown to the right of the front.<br><br>Fig. 133. H. | The same operation as in facing to the left of the rear. |
| To form an oblique line on the rear division, facing to the left of the rear—The front of the column thrown to the left of the rear.<br><br>Fig. 134. I. | The divisions of the column will countermarch, and it will then become a column with its left in front.<br>The head division will be placed with its right flank behind and fronting to the left of the new line—The rest of the column will march to the right, and place the right flanks of their divisions behind the new direction.<br>The line will be formed by divisions wheeling to the right. |
| To form an oblique line on the rear division, facing to the right of the front—The front of the column thrown to the right of the rear.<br><br>Fig. 134. K. | The divisions of the column will countermarch and become a column the left in front.<br>The head division will be placed with its right flank behind the new direction.<br>The rest of the column will march to the left, and place the right flanks of their divisions behind the new direction.<br>The line will be formed by divisions wheeling to the right. |
| To form an oblique line on the rear division, facing to the left of the front—The front of the column thrown to the right of the rear.<br><br>Fig. 134. L. | The same operation as in forming to the right of the front. |

| Nature of Movements. | Circumstances of Execution. |
|---|---|
| To form an oblique line on the rear division, facing to the right of the rear—The front of the column thrown to the left of the rear.<br><br>Fig. 134. M. | The column will countermarch and become a column with its left in front.<br>The head division will be placed with its right in the new line—The rest of the column will march to the right, and place their rights in the new line.<br>The line will be formed by wheeling to the right. |
| If the battalion is to form a square.<br><br>Fig. 135. | The center grand division will stand fast. The rest of the battalion will face towards it, and the head of each company will disengage to the rear. The companies will march and form in two open columns, behind the center grand division—The companies of the flank grand divisions will wheel outwards to form the flanks of the square; and the grenadier and light companies will close up and countermarch, if necessary to form the rear face of the square. |
| When the square marches. | The half company of the flank faces, will wheel backwards, and again place themselves in column. |
| When the square halts. | The half company of the flank faces, wheel up, and close the square. |
| From the square to form an oblong.<br><br>Fig. 136. | The left company of the center grand divisions will double behind the right company, and at half distance—The light company will also double behind the grenadiers, and the flank sides of the square will close in. |

From

| Nature of Movements. | Circumstances of Execution. |
|---|---|
| From the battalion in line to form an oblong. | In the same manner as the square is formed, except that the left company of the center grand division, doubles behind its right company. The other grand divisions, grenadiers, and light company, form in columns (behind the center grand division) of half companies; the flank faces are then formed by the wheeling up of the half companies. |
| From the square or oblong, to form in line on any given company.<br><br>Fig. 137. | Any one company is placed in any given direction, the other companies face towards their point in line, disengage their several heads if necessary, and are ordered to march in file to their points in battalion. |
| When from column of march, it is necessary to make front to both flanks.<br><br>Fig. 138. | The first division halts—The rest of the column will take half distance and halt. The half divisions will wheel outwards, halt, and form an oblong closed in the rear by the last division.<br>When there are several battalions in the column, each will form as above, closed by its own front and rear divisions. |
| When column of march is to be resumed. | The half divisions will wheel backwards into column, and the march will be resumed. |

When

| Nature of Movements. | Circumstances of Execution. |
|---|---|
| When the oblong is composed of more than one battalion.<br><br>Fig. 139. | Such number of central companies as are to form the front will be named—The rest of the body on each flank will place themselves in open column of companies behind it. As many of the last companies as are necessary will close the rear of the oblong. |
| When the oblong is to form with the sides six deep.<br><br>Fig. 140. | The companies will close to half distance when necessary—Wheel by half companies to the flanks, close up and form six deep—The front and rear faces will be strengtened by the grenadiers and light companies, which for that purpose are in the interior of the column or oblong. |
| When the oblong is to form in line.<br><br>Fig. 141. | The named company will be placed in the intended direction; the other companies will be formed and stand in column—The companies of the same battalion as the named company will file into line—The other battalions will march in column into line and then form. |
| In open ground which will allow of the movement of the line in front, and when a change of position is to be made on a flank, or on a central point of the line, by the marching up of divisions in front.<br><br>Fig. 42. | The given division is wheeled into the proposed direction. The other divisions of the line wheel separately an equal portion of the circle, so as to be parallel to the given one—Those that are behind it march up in front, and line anew with it; those that are before it face to the rear, march, line with the given division and then front. In some situations from open column, the divisions may thus by oblique marching move up into line. |

ECHELLON.

# ECHELLON.

| Nature of Movements. | Commands given by Leaders of ||  Circumstances of Execution. |
|---|---|---|---|
| | Battalions. | Divisions. | |
| When the battalion is marching, and the right wing is to form echellon to the rear. Fig. 79. | *The right wing of the battalion will form echellon to the rear on the 7th division at half distance to form oblique line.* | | A caution. |
| | | Halt | The fix right divisions. |
| | | Left march<br>Left incline<br>Forward | The sixth division, and successively every other. |
| | Halt | Halt | Line and echellon when ordered. |
| When halted, and that the echellon is ordered to form oblique line to the right. Fig. 80. | *The divisions of the echellon will wheel to the right into oblique line.* | | A caution. |
| | | Right wheel | Pivot man faces. |
| | | March | Each division wheels up. |
| | | Halt | In the oblique line. |

Commands given by Leaders of

| Nature of Movements. | Battalions. | Divisions. | Circumstances of Execution. |
|---|---|---|---|
| When the oblique line is to resume echellon. Fig. 81. | *The divisions will form in echellon* | | A caution. |
| | | *Left wheel* *March* *Halt* | Pivot man faces square. |
| When the echellon halted and formed to the rear is ordered to advance and form echellon to the front on the same division. Fig. 82. | *The echellon will be advanced and formed to the front.* | | A caution. |
| | | *March* | The whole divisions move forward. |
| | | *Halt* | Each division successively in its relative situation. |

# CLOSE COLUMN.

From LINE to form CLOSE COLUMN—The Right in Front.

Commands given by Leaders of

| Nature of Movements. | Battalions. | Divisions. | Circumstances of Execution. |
|---|---|---|---|
| | *The line will form in close column in front of the right division.* | | A caution. |

From

| Nature of Movements. | Commands given by Leaders of | | Circumstances of Execution. |
|---|---|---|---|
| | Battalions. | Divisions. | |

From line to form close column in front of the right division.

Fig. 45.

Battalions: *To the right face.*

Divisions: *March*

*Halt*

*Front*

*Left dress*

Circumstances of Execution:
- The right division stands fast, the rest face to it. The heads of files disengage to the left.
- And file into close column.
- The head of each division when at its point in column, and the rear closes in.
- By facing to the left.
- And cover on the pivot flank.

---

From line to form close column in front of the left division.

The same operation as above to the left.

---

From line to form close column behind the right division.

Fig. 45.

Battalions:
*The line will form close column behind the right division.*

*To the right face*

*March*

Divisions:

*Halt*

*Front*

*Left dress*

Circumstances of Execution:
- A caution.
- The right division stands fast; the rest face to it. The heads of files disengage to the right.
- File into close column.
- The head of each division when at its point in column, and the rear closes in.
- By facing to the left.
- And cover on the pivot flank.

|  | Commands given by Leaders of | | |
|---|---|---|---|
| Nature of Movements. | Battalions. | Divisions. | Circumstances of Execution. |
| From line to form close column behind the left division. | | | The same operation as above, to the left. |
| From line to form close column on a central division, the right in front. Fig. 45. | *Form close column on the 6th division of the battalion the right in front.* | | A caution. |
| | *Inwards face* | | The sixth division stands fast.—Those on the right of it face to the left—Those on the left face to the right. The heads of files disengage to the right. |
| | *March* | | The divisions on the right to place themselves in front. The divisions on the left to place themselves in rear of the named division. |
| | | Halt | Each division halts, and closes in when it arrives at its place in column. |
| | | Front | By facing to the proper front. |
| | | Left Dress | And cover on the pivot flanks. |
| From line to form close column on a central division, the left in front. | | | The same operation as above, to the left. |

To

( 45 )

| Nature of Movements. | Commands given by Leaders of ||  Circumstances of Execution. |
|---|---|---|---|
| | Battalions. | Divisions. | |
| To form several close columns from part of the same line. Fig. 10. | | | The parts of the line which compose each column are named. The division on which each column is to form is also named.—The several columns then separately are ordered to form. |

From CLOSE COLUMN, the Right in Front—To Form LINE.

| Nature of Movements. | Commands given by Leaders of || Circumstances of Execution. |
|---|---|---|---|
| | Battalions. | Divisions. | |
| When from column of march of platoons the battalion forms close column to deploy. Fig. 142. | | *Move up to quarter distance* | When approaching the alignement, the head halts or shortens the step, and the rear platoons are ordered to close up quick to quarter distances. |
| | | | The column resumes its ordinary march. |
| | | *Halt* *Close distance* | When the head arrives on the alignement it halts, and the rear moves up close. |
| | *Form divisions* | *Face* *Halt* *Front* *Eyes, &c.* *March* *Halt* | Divisions are then formed by the second platoon doubling up to the right or left (as proper) of its head one; and thus every other platoon. An officer is on the flank of each division. |
| | *Divisions close up* | *Divisions close up* *March* *Halt* | The whole are again ordered to close to the front, and the column remains ready to extend into line. |

From

( 46 )

| Nature of Movements. | Commands given by Leaders of ||| Circumstances of Execution. |
| | Battalions. | Divisions. ||
|---|---|---|---|
| From battalion column to form in line on the front division.<br><br>Fig. 46. | *The line will form on the front division* | | A caution.—The column being supposed placed on the alignment. |
| | *To the left face* | | The front division stands fast, the others face to the left. Non-commissioned officers are ready to move out to shew the ground. The conducting officers are on the left of each division. |
| | *March* | *March* | To the left parallel to the new line. The heads of files being dressed. |
| | | *Halt*<br>*Front*<br>*Eyes to the right* | Successively each division (2, 3, 4, 5, &c.) from right to left as it comes opposite its point in line. The rear of the file closing in if necessary. |
| | | *March* | By the flank next the division of formation and directed by the right officer. |
| | | *Halt*<br>*Dress* | Each as it arrives in the new line. |
| | | | And thus successively division after division from the right to the left, till the battalion is formed. |
| | *The line will form on the rear division* | | A caution.—The rear division sends two non-commissioned officers to place themselves immediately before each flank of the front division, exactly to mark the ground to which it afterwards is to move up; and the leader of the front division takes his just |

| Nature of Movements. | Commands given by Leaders of | | Circumstances of Execution. |
| --- | --- | --- | --- |
| | Battalions. | Divisions. | |
| From battalion column to form in line on the rear division. Fig. 47. | | | juft points of march to the right, which are exactly afcertained by the adjutant, or by his fergeant, who runs along the line, and marks the point where his flank is to be placed. |
| | | *To the right face* | The rear divifion ftands faft, the others face to the right. The conducting officers are on the right of divifions. |
| | | *March* | The whole move except the rear divifion. The front divifion files to the right, exactly along the line of formation, the others move parallel behind it, their feveral heads dreffing. |
| | | *March* *Halt* *Drefs* (The rear divifion only.) | When the rear divifion is uncovered, it alfo receives the word *march*, moves forward and halts at the non-commiffioned officers, who mark its place in the new line. |
| | | *Halt* *Front* *Eyes to the left* | When the divifion immediately preceding the rear one has marched fufficiently to the right fo as to be oppofite its ground in line, it halts—fronts—dreffes to the left. |
| | | *March* *Halt* *Drefs* | When uncovered it marches by command from the left officer and halts in line, with the divifion of formation which has arrived at its ground. |
| | | | And thus fucceffively divifion after divifion from left to right of the battalion till it is completely formed. When the whole, if thought proper, may be ordered to drefs to the right. |

From

( 48 )

| Nature of Movements. | Commands given by Leaders of | | Circumstances of Execution. |
|---|---|---|---|
| | Battalions. | Divisions. | |
| From battalion column to form in line on the center division.<br><br>Fig. 48. | *The line will form on the 5th division of the battalion.* | | A caution.—The named division sends two non-commissioned officers to the front to mark the ground where it is to arrive.—The leader of the front division observes exactly the points he is to march upon, which are marked by the adjutant, sergeant, &c. |
| | *Outwards face* | | The named division stands fast. Those in front face to the right—Those in rear face to the left. |
| | *March* | | The front divisions as directed in fig. 46. The rear divisions as directed in fig. 47. The named division remaining halted. |
| | | *March* ⎫<br>*Halt* ⎬ The named division only.<br>*Dress* ⎭ | When the named division is uncovered, it moves forward to its place in line and halts at the two non-commissioned officers advanced from it. |
| | | *Halt*<br>*Front*<br>*Eyes to the left*<br>*March*<br>*Halt*<br>*Dress* | Successively the front divisions as directed in forming line on the rear division. |
| | | *Halt*<br>*Front*<br>*Eyes to the right*<br>*March*<br>*Halt*<br>*Dress* | Successively the rear divisions as directed in forming line on the front division. |

When

| Nature of Movements. | Commands given by Leaders of | | Circumstances of Execution. |
|---|---|---|---|
| | Battalions. | Divisions. | |

| Nature of Movements. | Battalions. | Divisions. | Circumstances of Execution. |
|---|---|---|---|
| When the close column halted is to take a new direction—To the left.<br><br>Fig. 143. | *To the left* <br> *face* <br> *March* | *Halt* <br> *Front* <br> *Dress* | The front division of the column will be placed in the new direction.<br>All the other divisions of the column.<br>Each division separately, when it arrives at its point, and covers in column. |
| When the column is to make front to its rear by countermarching.<br><br>Fig. 144. | | | The divisions of the column will each separately countermarch in same manner as directed for the open column—The divisions are supposed at a sufficient distance to allow of this operation. |
| When a column of march of several battalions are to form in close column, and from thence extend into line.<br><br>Fig. 49. | | | The column marching by platoons closes to the front to quarter distance—The whole halt when near the new line.<br>The several battalions disengage from a named one, and stand in an echellon position.<br>The named battalion, as well as all the others, advance, and halt on the given line of formation, with an interval of two platoons betwixt each.<br>Divisions are formed by every other platoon doubling up to the left of the one preceding it, and the rear divisions move up close.<br>The division on which the line is to be formed is named—The rest of the divisions and columns face outwards from it, march to the flanks, and successively deploy into line, as they arrive at their several points of formation. |

( 50 )

| | |
|---|---|
| If several close columns are halted at accidental distances, but with their heads dressed, and are ordered to form in one line.<br><br>Fig. 49, 52, 68. | At whatever distance the heads of close columns are halted from each other, the separate battalions will first move up into line—The point and division on which the whole are to form will then be named.—The whole will extend from it.—The distances and commencement of movement will be taken from the named point, so that the outward battalions may move successively, as it becomes necessary to preserve their distances from the inward ones.—The flank points of halting and forming for each battalion will be ascertained by the adjutants, from the named one, as soon as possible. |

| Nature of Movements. | Commands given by Leaders of ||Circumstances of Execution. |
|---|---|---|---|
| | Battalions. | Divisions. | |
| Fig. 49. | *The column will close to the front to quarter distance* | *Take quarter distance* | The front of the column either halts or shortens its step, and the rear platoons close up quick to quarter distance; an interval equal to the front of a platoon is reserved betwixt each battalion, and the column when closed resumes its ordinary march. |
| | *The column will halt* | *Halt* | When arrived within about 200 yards of where the line is to be formed—The head halts, and all distances are closed up to the front. |
| | *The battalions will form echellon on the 3d battalion* | | The third, or any other battalion is thus named, as that on which the column is to deploy, and the points in the alignement are already marked out. |
| | *Battalion Face* | | The third battalion stands fast, each other one is separately ordered to face.—Those in front of the third battalion to the right, those in rear of it to the left. |

When

( 51 )

| Nature of Movements. | Commands given by Leaders of ||| Circumstances of Execution. |
| --- | --- | --- | --- |
| | Battalions. | Divisions. | |
| When a column of march of five battalions forms close column, and then extends into line. Fig. 49. | | *March* | Each marches quick to the flank, without opening out. |
| | | *March* *Halt* *Dress* | (The named batt. only.) When the third battalion is uncovered it moves forward to its place in line, and halts at its given points. |
| | | *Halt* *Front* *Eyes to the March* *Halt* *Dress* | The battalions that are marching to the flanks, as soon as they have acquired an interval of two platoons from each other, will successively halt, front, dress inwards, march, halt, and dress with the third battalion which has advanced and halted. |
| | *The platoons will form divisions* | *Left face* *March* *Halt* *Front* *Eyes to the right* *March* *Halt* *Dress* | The battalions being thus placed on the alignement with the above intervals in column of platoons at quarter distance, are ordered to double up, each other platoon to the left, to form divisions. |
| | *Form close columns* | *Close to the front* *March* *Halt* | The whole close up to the head of their respective battalions. |
| | *The line will be formed on the 5th division of the 3d battalion* | | A caution.—Points, &c. being all ascertained. |
| | *Outwards face* | | The two right battalions and the front divisions of the third to the right, all the rest to the left. |

Tt 2

Fig.

| Nature of Movements. | Commands given by Leaders of | | Circumstances of Execution. |
|---|---|---|---|
| | Battalions. | Divisions. | |
| Fig. 49. | | *March.* | The whole move to the flanks except the named division, which advances into the alignment, and the third battalion proceeds to make a central formation on it. |
| | *Halt Front Form in line, &c.* | | The other battalions continue their march till they arrive at the points where their inward flanks are to be placed, they then successively halt, front, and deploy on whatever division of each (head or rear), is to be there fixed when in line. The other attentions already ordered, must be exactly observed. |
| When two columns are to form in line in any given position. Fig. 51, 53. | | | Points are prepared—the colums close up—halt—disengage the battalions—march into the new line—the division or divisions of formation are named, and all the others relatively deploy into line. Or, when the several battalions are disengaged from the general column, their heads may be directed to their several prepared points, where they take distances, wheel into line, and form. |
| Formation from columns in two lines. Fig. 52. | | | If there are two columns, composed each of part of two lines—The battalions of the second line will halt at a proper distance from the first, and deploy or form in line, in the same manner as the first does. |

Formation.

| Nature of Movements. | Circumstances of Execution. |
|---|---|
| Formation in one line from several columns.<br><br>Fig. 54. | When more than two columns are to form in the same line—The intermediate distances betwixt their heads when halted, will be filled up by the whole or parts of the interior columns as is seen necessary, and every thing else will deploy outwards. |
| When the close column halted is to form in line to the left, in the prolongation of that flank, and on the rear division, by taking distances.<br><br>Fig. 145. | The rear division remains halted—The others in front of it move on, and halt division after division from the rear, as they get their wheeling distances—When the whole or part is thus in open column, the divisions wheel up and form the line. |
| When the close column halted is to form to the left, in the prolongation of that flank, and on the front division, by taking open distances.<br><br>Fig. 146. | The front division remains halted—the others face to the right about—march, and halt separately as they acquire wheeling distances from their then rear—Face to the right about, and when the whole or part is thus in open column, the divisions wheel up and form in line. |
| When the close column halted is to form in line to the left, in the prolongation of that flank, and on a central division, by taking open distances.<br><br>Fig. 147. | The central division is named—The divisions in rear of it face to the right about—The whole then march and halt successively, as they acquire wheeling distances from the central divisions—The rear divisions again face to the right about, and when the whole or part is thus in open column, the divisions wheel up and form in line. |

| Nature of Movements. | Circumstances of Execution. |
|---|---|
| When the close column halted is to form in line to the right, by deploying in the prolongation of that flank, or obliquely on any given division.<br><br>Fig. 148. | The head division will be placed in the new direction—The named division and all such as are in front of it will then file and cover it in such new direction—The rear divisions of the column remain halted.<br>The front divisions will then deploy to the right on the named division; and those in the rear will march on and wheel separately to the right, each as it comes opposite to its ground, and march into the new line. |
| When the close column halted is to form in line to the left, by deploying in the prolongation of that flank, or obliquely on any given division.<br><br>Fig. 149. | The front divisions will be wheeled to the left, placed, and deploy to the right as before.<br>The rear divisions will file to the left in mass into the new line—They will then face to the right about—march on in the new direction—take wheeling distances from their then rear—face separately to the right about, and form in line by wheeling to the left. |
| Fig. 143. | If on the above occasions, the whole close column can be first placed in the new direction, it will then as usual deploy into line. |
| When the battalion is to form in double column of attack from the center.<br><br>Fig. 151. | The two center companies stand fast, the others in each wing face inwards, disengage, file to the rear, and place themselves in column at half or quarter distance, as ordered—The grenadiers taking the front, and the light infantry the rear of the column. |

| Nature of Movements. | Circumstances of Execution. |
|---|---|
| Double column of several battalions.<br><br>Fig. 152. | If the column of attack is composed of several battalions, they may draw up behind each other—Then separately form as above, and in this manner advance. Should the front of the column of attack be increased, the depth will be proportionally diminished. |
| When the double column is to extend into line.<br><br>Fig. 153. | The rear companies deploy and line with the center, either by filing, or in some cases by a sharp incline, where it can be done. |
| When the battalions of the second line are to form in double column on the center companies.<br><br>Fig. 154. | They form in same manner as the double column of attack—The divisions are at half distance, and the grenadiers and light infantry may be in the rear—In that situation they are ready to deploy into line, or to allow a passage to the first line if it retires, or to take up any ordered position, or to form the oblong if necessary. |
| When the companies of the second line double, to allow the first line to pass through.<br><br>Fig. 155. | If the first line is obliged to retire—The left companies of the second line may be timeously ordered to double behind their right ones, will allow the first line to pass through, and will then resume their places in line. |

When

| Nature of Movements. | Circumstances of Execution. |
|---|---|
| When a line is to retreat in several squares or oblongs.<br><br>Fig. 156. | The several brigades may naturally and separately compose the squares or oblongs—The central companies which form the front of each will be named, the others will be placed in open column from each flank behind them—The grenadiers and light company will be in the interior, to be applied according to circumstances—The companies in column will be at half or whole distance as is thought necessary; some one body will be named as the regulating one, and in this order will the whole retire, and be prepared at any instant to form into line. |
| When to retire part in line, part in column.<br><br>Fig. 157. | Such battalions or portions of the line as are ordered, may form columns behind their given divisions—In this order the whole will retire, partly in line, partly in column, sufficient distances being preserved for the columns to extend into line when thought necessary. |
| When the solid column is to make front on all sides.<br><br>Fig. 158. | When the column is closed up, the two front and two rear divisions will face outwards—The interior division will also face outwards, dress the head of their flank files; and the drums, music, &c. will find place in the center of the column.<br>The five first ranks all round the column will face outwards—The two first will kneel, and present bayonets—The three next ranks will fire separately by ranks, if necessary to repulse the attack of cavalry. |

The general PRINCIPLES that should direct the movements of all bodies, great or small, having been fully detailed; the following application of them to the formal parts of a REVIEW or INSPECTION, is meant not (unnecessarily) to vary from the customary practice, and is therefore given as an example of what may be expected on that occasion from one or more battalions. Such particular movements are also superadded as are simple, essential, and sufficient for taking up any position in the field, which may be required from the battalion or line in any situation.

ORDER of REVIEW for a battalion of infantry, consisting of ten companies.

*Receiving the general.*

The BATTALION assembles on its parade, is formed, and there told off—Marches to its review ground, in column of companies or platoons, and there again forms in close order.

Before the reviewing general arrives, firelocks are shouldered, bayonets are fixed, and the battalion takes open order as directed in the formation of the battalion.

In this disposition, and the whole dressed to the right, the general is awaited. He is to be received with the compliments due to his rank, as set forth in the regulation of military honours.

U u                                                      Com-

| Given by | Words of Command. | |
|---|---|---|
| Commanding officer. | *Present your arms* | When he approaches the right of the regiment, the word of command is given; and the instant the men have performed the last motion of the rest, the musicians and fifers begin to play, and all the drummers to beat. |
| | | The officers are to salute singly—each one endeavours to time the first, in such manner, as just to have finished the last motion when the general is opposite to him, viz.    *Standing salute.* |

      Staff, who take off their hats.
      Major.
      Grenadier officers.
      Battalion officers of the right wing.
      Colonel, lieut. colonel (and colours if proper).
      Battalion officers of the left wing.
      Light company officers.
      Adjutant.

| Given by | Words of Command. | |
|---|---|---|
| Commanding officer. | *To the left, face* | When the reviewing general is arrived on the left of the battalion, and turns in order to pass along the left flank, the command is given; on which the regiment faces to the left.   *Facings of the battalion.* |
| | *To the left, face* | As soon as the general has passed the left flank of the regiment, and is in the rear, the command is given; on which the regiment again faces to the left. |
| | *To the left, face* | The same word is repeated when the general, after having passed along the rear rank, comes upon the right flank of the regiment. |
| | *To the left, face* | The same a fourth time also, which brings the regiment to its proper front, after the general has gone quite round it. |

During

| Given by | Words of Command. | |
|---|---|---|
| | | During this operation of parade, and in order to add to appearance—The sergeants may be directed to fall back and form in a fourth rank in the rear at four paces distance; and in such case they will perform the same motions with their halberts and fusils in the facing, as their respective officers do. *Sergeants.* |
| Commanding officer. | *Shoulder firelocks Rear ranks close March* | While the general is proceeding to place himself in the front—The words of command are given. |
| Commanding officer. Leaders of divisions. | *By grand divisions to the right wheel March Halt* | The battalion is now formed in close order, and prepared to march past the general.<br>The grenadiers, the four grand divisions, and the light company, wheel separately a quarter of the circle to the right, and halt.<br>The colonel moves to the head of the first grand division.<br>The lieut. colonel to the rear of the fourth grand division.<br>The major to the head of the grenadier company.<br>The adjutant to the rear of the light company.<br>Staff officers do not march past.<br>Musicians are four paces before the colonel.<br>Pioneers place themselves in a single rank ten paces before the grenadiers.<br>Drummers and fifers are in the rear of their respective divisions. |

N. B. The march by grand divisions is customary, otherwise that by companies would be prescribed, and the desirable circumstance of equality of divisions in the column thereby preserved.

Post of field officers, &c. in order to march past.

| Given by | Words of Command. | | |
|---|---|---|---|
| | | As the grenadiers and light company are out-flanked after the wheel by the grand divisions, they will shift to the left, that the pivot flanks of the whole may cover. | |
| Commanding officer. | *March* | { The whole column is now put in march; at this word of command, dressing is to the left, the proper pivot flank. | March of the column. |
| Leaders of divisions. | *To the left—wheel* <br> *Halt* <br> *March* | { The grenadier company when it has marched sufficiently to the flank, and then each grand division and the light company in succession, as they come up to the ground on which the grenadiers wheeled. | |
| | *To the left—wheel* <br> *Halt* <br> *March* | { The grenadiers, grand divisions and light company successively, when they come near to the line in which they are to march past the general. | |

In this second wheel, the leading grand division will observe to begin its wheel, as much short of the ground on which the grenadiers wheeled, as will at the conclusion of it, bring all the right flanks to line instead of the left ones: But the light company will move on to the exact ground of the grenadiers before they wheel.

After the front division has made this wheel, should the column have been marching in quick time, it will be halted, and march off in slow time.

| Given by | Commands. | | |
|---|---|---|---|
| | | When the divisions have arrived on the line in which they are to march past the general, and are within fifty or sixty paces of him, they will successively prepare to pass in parade. | Marching past the general. |
| Leaders of divisions. | *Eyes to the right* | On which the division dresses to the right, being before this dressed to the left. | |
| | *Rear ranks take open distance* | The rear ranks take their open distance, by allowing the front rank to move on—drummers shift to the left flank of their divisions—officers move out three paces to the front—serjeants move into the places of officers in the ranks. | |
| | | In marcing at open ranks the leading officer of each is responsible for the proper distance being kept from the front rank of the division preceding him. His own front rank must invariably preserve its fixed distance of three paces from him. The rank of officers dress to him; eyes are turned a little to the right, and they divide the ground in order to cover the front of their divisions. | |
| | | The battalion officers when they arrive at a proper distance from the general, must prepare to salute, viz. | Marching salute. |
| | | When opposite the general. | Major. |
| | | When within six paces of the general, and trail their fufils when ten paces past him. | Grenadier officers. |
| | | When he has saluted at the head of the battalion, places himself near the general. | Colonel. |
| | | When within six paces of the general, and recover swords when ten paces past him. | Battalion officers. |

Same

| Given by | Commands. | | |
|---|---|---|---|
| | | Same as battalion officers. | Lieutenant colonel. |
| | | Same as the grenadier officers. | Light company. |
| | | Same as the major. | Adjutant. |
| | | After passing the general, they draw up opposite to him, and continue playing till the regiment is past. | Music. |

Leaders of divisions. — *Rear ranks close to the front.*

{ After having passed the general, the officers commanding divisions will successively close the rear ranks, before they make the first wheel. Those at the head of the leading ones taking care each in his turn, not to give the word, till the division next succeeding has marched past the general. When the ranks close up, each individual is to resume the same post he held when the column was first put in march.

When the rear ranks close, or open on the march—In the one case they will run up, in the other they will slacken their pace, till the due distance is attained; in both cases the front rank continues to proceed at its then rate of march.

Leaders of divisions. — *To the left—wheel. Halt. March.*

{ Each division successively when opposite the ground where the left of the regiment stood.

The leading division must take care not to march too fast, till the last division has passed the general, and closed ranks.—When that is done, column of companies will then be formed in order to march past in quick time; if such operation is thought proper.

| Given by | Commands. | | |
|---|---|---|---|
| | *Left companies.* *Double* | The left companies of the grand divisions allow the right ones to advance clear of them, they then incline to the right, cover the right ones, and divide the former interval. | |
| | *Companies to the left.* *Dress and cover* | The column is now ordered to dress to the left, the proper pivot flank, and officers post themselves accordingly. | |
| | *Halt* | Repeated successively from the front division to the rear. | |
| Leaders of divisions. | *Quick. March* | The officers commanding all the divisions from front to rear, immediately repeat the same, and the whole move on in quick time (gaining any distance that may have been lost in marching past) till the front arrives on the left of the original ground. | Marching past in quick time. |
| | *To the left—wheel* *Halt* *March* | Leading divisions followed by the others successively, and the column proceeds along its original line. | |
| | *To the left—wheel* *Halt* *March* | When the front division has arrived at the point where the right of the battalion first stood, it wheels, and is followed by the rest of the column. | |
| | *To the left—wheel* *Halt* *March* | When the front division arrives on the line in which it is to pass the general. | |
| | *Right—dress* | Each division forty or fifty yards before passing. If firelocks have been sloped, they are ordered to be carried. | |

March past the general in quick time. Officers do not salute, or pay any compliment, but are attentive to preserve the proper intervals betwixt the divisions—The major still leads the grenadiers—The colonel the battalion—The adjutant in rear of the light company—The colours in front of the fourth company—The music in front of the grenadiers

Pass in quick time.

| Given by | Commands. | |
|---|---|---|
| | | diers—The pioneers in front of the whole—The drummers on the left of their companies. |
| Leaders of divisions. | *Eyes to the left* | The several companies after passing will successively dress to the left, the proper pivot flank. |
| | *To the left—wheel* *Halt* *March* | The column will continue its march. The head division will make its two wheels to the left, followed successively by each other one, and march with just distances along the line, on which it is to form. |

| | | | |
|---|---|---|---|
| Commanding officer. | *Halt* | When the grenadiers are arrived at their ground, the major halts them, and each other division will halt instantly by order of its own leader, at its just distance. | Forming the battalion. |
| | | The major now carefully corrects the pivot flanks in an exact line previous to forming. | |
| | *To the left wheel and form battalion* *March* *Halt* *Right—dress* | The companies wheel to the left together. Halt when in the line of the several standing flanks, and immediately dress to the right. But such operation of dressing only regards the internal parts of the company, for the pivot men, who before the forming in line were placed critically true, must remain immovable. At the foregoing word march, every individual moves to his proper place in close order, in which situation the battalion is now halted. | |
| Commanding officer. | *Rear ranks, take open distance* *March* | Open order is now taken. | |
| | *Present your arms* *Shoulder* | General salute. | |

The

| Given by | Commands. | |
|---|---|---|
| | | The detail of each motion in the different salutes, and positions of the sword and fusil, are according to regulation. |
| Major. | *Manual* | The major comes forward in front of the center—Unfixes bayonets—proceeds with the manual as directed by regulation.<br><br>After the manual – The battalion is again formed in close order, and the major returns to his post in the rear. |
| Commanding officer. | *Prime and load* | The commanding officer orders the battalion to prime and load.<br><br>The regiment is now prepared for such firings, movements or manœuvres mixed with firings, as the general or commanding officer shall determine, and are consonant to the principles established by regulation.<br><br>Among others the following are pointed out, to be varied, to follow, and to be applied according to circumstances of ground, and the intention of the commanding officer. |

By
{ Grand divisions
Companies
Platoons } from { Flanks to center
or
Center to flanks.
Wings
Battalions } to the { Front.
Oblique.
Files from the right of companies.

These firings seem to comprehend all the variety that is neceſſary. The execution of them muſt be particularly detailed, and they will be applied as thought proper to the movements, and ſituation of the battalion.

## MOVEMENTS and MANOEUVRES of the Battalion.

The following manœuvres are ſuch as include the various movements that can be required from a battalion; the detail of their execution is pointed out under the articles of the line, the column, and the echellon—They are here arranged as following each other, but they may be any otherwiſe claſſed or ariſe from different ſituations, by adding or altering the connecting circumſtances.

*Firings.*    The firings may be applied to moſt of thoſe movements as ſhall be thought proper.

*Filings.*    The filings and breakings of the battalion by companies, are in general the moſt eligible, as the men receive commands from the officers, to whom they are moſt accuſtomed.

*Commanding officers.*    The commanding officer mounted is in front to lead and determine the movements, and will fall into the rear, or diſmount while the battalion is firing.—The other mounted officers will aſſiſt in directing the march, correcting diſtances, and dreſſing the pivot flanks of the column in all ſituations.

*Light company.*    The light company and grenadiers are ſuppoſed acting in line with the battalion.

The light company may be occasionally placed, half of it behind each flank of the battalion—In that situation it is ready to cover the front, rear, or flanks of the column when in march; to protect the forming of the line, or to cover its retreat—For these purposes it will be detached from time to time, and act in division or individually as circumstances require.

---

1.—The battalion will move forward.
    Halt.
    Fire.

*Advancing.*

*Fig. 162.*

---

2.—The battalion will retire.
    Front.
    Fire.

*Retiring.*

---

3.—The battalion will wheel to the right, into open column of company or platoons.
    March in column.
    Halt.
Form in line by wheeling up to the left.

*From open column forming line to the flank.*

*Fig. 163.*

---

4.—The battalion will form in close column of companies the left in front, and behind the left company.
    Face, to the left.
    March, to the flank.
    Halt.
    Front.
Form the line on the 2d battalion company.

*Forming of close column and line.*

*Fig. 164.*

---

5.—The battalion will march from the right to the front in open column of companies.

From column of march forming line to the front.

Fig. 165.

The leading company will wheel on a new direction oblique to the right, and halt when the 3d company has entered it.

The rear companies will then file to the left, and place themselves in the new column.

The line will be formed oblique with respect to its original direction, by wheels of the quarter circle to the left.

---

From column of march forming line to the rear.

Fig. 166.

6.—The battalion will march from the left, to the rear in open column of companies.

The leading company will wheel to the right and halt.

The rear companies will then file to the left, and place themselves in the new column.

The line will be formed, by wheels of the quarter circle to the right.

N. B. If the line is to be formed fronting to the rear—The leading company of the column will wheel to the left; when it halts the others will file to the right, and the line be formed by wheels of the companies to the right.

---

From column of march forming line on a central division.

Fig. 167.

7.—The battalion will march from the left, to the front in open column of companies.

The column will halt.

The 3d company will wheel to the left perpendicular to a new direction.

The front divisions will file to the left and rear, the rear divisions to the right and front, and place themselves a-new in column.

The line will be formed by wheeling to the right.

---

From column of march forming close column and line.

Fig. 168.

8.—The battalion will form column at half intervals, the left in front of the 4th company.

The companies will counter-march.
The column retires.
Halts.
The companies will counter-march.
The column will march to the right in file.
Halt.
Front.

Form

Form clofe column to the front divifion.
Form in line on any divifion ordered.

---

9.—The battalion will wheel by divifions to the left.
    Face to the right.
    March in file to the front.
    Halt.
    Face to the left.
    Wheel to the right into line.

*The divifions of the column filing to the front and forming in line.*

N. B. In the courfe of the march the divifions if ordered may fall into fingle file, each center, and rear rank man doubling behind his front man, and again moving up to their compleat file, before the column halts to form in line.

*Fig. 169.*

There may be occafions where the divifions marching in file to the front will halt, and the line be formed, by each man at once moving up to his neighbour.

---

10.—The battalion will wheel by divifions to the right.
    Face, to the right ⎫
    File, to the rear   ⎬ The divifions may fall into fingle file, as before mentioned.
    Halt              ⎭
    Face to the left.
    Wheel to the left into line.

*The divifions of the columnfiling to the rear and forming in line.*

*Fig. 170.*

N. B. The right is the leading divifion from which diftances are kept, and with which heads of files fhould line.

---

11.—The battalion will form clofe column on any named company, the right in front.
    The front company will wheel into a new direction.
    The rear companies will file, and cover it.
    The line will be formed on any named company.

*The clofe column taking a new direction and forming in line.*

*Fig 171.*

---

12.—The companies will each wheel $\frac{1}{8}$—$\frac{1}{16}$ or fuch part of the circle as will bring them parallel to the original front, and place them in echellon pofition.

Face

( 70 )

**From position in echellon, the battalion dresses to the rear division.**

Fig. 172.

Face to the right about.

March ⎫
Dreſs ⎬ To the company moſt retired, and which does not face
Front ⎭ or move.

---

**The companies retire from the left in echellon, and form in line on a center division.**

Fig. 173.

13.—The right company will ſtand faſt, the others go to the right about, retire; and each faces when at half diſtance from the one to the right of it. The whole in echellon poſition.

The 4th company will ſtand faſt and the others will dreſs in line to it—By the left advancing and the right retiring and fronting.

---

**The echellon formed from each flank to the rear.**

Fig. 174.

14.—The 3d company ſtands faſt—The others retire from right and left beyond each other, halt, and front at half diſtance.

The whole retires.
Halts.
Fronts.

The right companies in echellon cover to the left, each the 4th file of the one preceding it.

The left companies advance into line with the 3d company, and the right companies wheel to the right, and cover that flank.

---

**The echellon advanced, formed from the right, and an oblique poſition taken.**

Fig. 175.

15.—The right companies wheel up to the left.

The whole advances ⎫
The whole halts     ⎬ Part in line, part in Echellon.

The right companies move on and paſs each other, forming echellon advanced to the front at half diſtance, and the right file of each covering the 4th of the one preceding it.

The right companies wheel to the left, and form oblique on that flank.

---

16.—The left companies wheel each a ſmall part of the circle to the left, and halt.

March.

March.
Line with the right ones, and front.  *The line formed.*

---

17.—The alternate and left companies of the battalion retire one hundred yards and halt, making a degree of wheel on the left company, and also inclining during the movement to give the line a new direction.  *Retreat by alternate companies.*
The right companies retire two hundred yards in same manner, and front.  *Fig. 176.*
The left companies retire and join in line.

---

18.—The battalion advances—halts—fires a volley.  *The battalion advances, fires, and attacks with bayonets.*
The battalion advances—halts—fires a volley.
The battalion advances—halts—fires a volley.
The battalion advances—charges bayonets—halts.

---

SHOULDER.

OPEN RANKS.

GENERAL SALUTE.

---

WHEN several BATTALIONS are REVIEWED, or ACT together in LINE.

---

They will be individually formed, as already directed, with an interval of twelve yards betwixt each.

Two pieces of cannon are placed in the interval on the right of each battalion.

**Formation of the line.**

The light companies will compofe part of the general line, or may be formed in fecond line either feparate or collected—The fecond line will commonly be about two hundred and fifty yards from the firft.

If there are cavalry on the flanks, the neareft fquadron will be fifty paces removed—The front rank will drefs with the rear rank of the infantry when in clofe order, and with the center rank when in order of parade.

---

**March in column.**

The general will be received by each battalion feparately, and by order of its own commanding officer.

The whole line will wheel to the right into column, by fuch divifions as are ordered.

The column in marching paft, will occupy no greater extent of ground than when it originally wheeled. Its order is never to be broke or lengthned out, but when the chief commander fo directs. No particular battalion or the artillery, are allowed to increafe the diftances for their own partial appearance—The battalion guns will march two abreaft—The head of the column will take care not to hurry the rear—When in motion, the front of the column may be diminifhed if neceffary, and it may in like manner change its time of march.

---

**Forming in line.**

The leading divifion after paffing the general one hundred yards, may wheel to the left and purfue its march. Such of the rear battalions as find it neceffary, will after paffing, and at the proper time, break from the column of march, and point to their feveral places in the original line.

The battalions as they arrive at their original ground, will drefs the flanks of their divifions, wheel up to the left, and form fucceffively into line.

---

**The manual.**

The manual (when ordered) may be performed by command from the eldeft major; each battalion will take the time from its own advanced flugel men. And the feveral flugel men will take their motions from a center one.

The officers of the line will go to the rear by one word of command, taken from the right.

After

( 73 )

After the manual, rear ranks will cloſe, and officers take their poſts in battalion ſeparately, by one word of command.

---

Except the manual, all other operations or movements of the line, will be performed by the battalions ſucceſſively.

In all ſituations, ſome one particular battalion is named to regulate the whole as to pace and direction; each battalion in line moves by its own center; and the leaders conform to that of direction, which is generally a flank one. General attentions.

The commanding officer of the line or column, ſhould always addreſs his orders to the commander of the regulating battalion of the line, or head of the column; what he directs, the others will immediately repeat and take up from him.

The movements and halts of battalions and diviſions, though ſucceſſive, ſhould and may be almoſt inſtantaneous, if officers are quick, obſerving, firm, and decided in their words of command.

The general firings ſhould be executed ſeparately by each battalion.

---

The manœuvres of the line are ſimilar to, and derived from the ſame general principles, as thoſe of the ſingle battalion; they may be compounded, varied, and applied according to the ground, other circumſtances, and the intentions of the commanding officer. The greater the body, the fewer and the more ſimple ought to be the manœuvres required of it.

When the line breaks into ſeveral columns, diſtances and dreſſing are taken from one of the outward columns. General movements.

No conſiderable body ſhould ever be formed without a proportion of it being placed in reſerve or ſecond line; and more or leſs ſtrong, according to circumſtances.

The movements of ſuch ſecond line or reſerve, will always correſpond to that of the firſt line.

The cannon, whether in line or column, will accompany the movements of their proper battalions.

( 74 )

ORDER of REVIEW for a BODY of TROOPS, confifting of feveral BATTALIONS and SQUADRONS—For example, fix Battalions, and four fquadrons.

*The line formed.*

1.—The fix battalions are formed in one line, and divided into two brigades.

Two pieces of cannon are on the right of each battalion.

Two fquadrons of cavalry are on each flank.

The light companies are formed in fecond line, each one hundred and fifty paces in rear of its own battalion.

The whole are in order of parade, in the fame fituation in which they afterwards receive the general.

*Each regiment forms in column.*

*Fig. 177.*

2.—The right troops, and companies of regiments and the cannon remain on their ground.

Each regiment forms in folid column of companies and troops behind its own right.

Each light company advances one hundred and fifty paces in front of its battalion, and halts.

The whole remain in this order to wait the approach of the general.

*The line is formed from column to receive the general.*

3.—When the general (fuppofing him entitled to the falute of cannon) arrives oppofite the left, or center, the firft gun is ordered to fire; and there will be an interval of fix feconds betwixt each gun.

At the firft gun, each regiment by command, deploys into line, the light companies retire through the intervals to their pofts in 2d line—Rear ranks are then opened, and the whole compleatly formed in order of parade.

N. B. All this may be quickly done—And the cannon will have finifhed the falute of twenty-one guns before he approaches the right.

4.—When

4.—When the general is going along the front of the line, the mufic of the whole will play, and the trumpets and drums of the regiment he is paffing will found, and beat—The officers of each feparately faluting.

*Receiving the general.*

N. B. During this operation the whole of the light infantry will have clofed into the right company, formed in one battalion and there receive the general.

---

5.—When the general is on his return to place himfelf oppofite the center—The line will take clofe order, and break to the right into column of troops and companies.

*The line forms column of march.*

---

6.—When the general is near his pofition, the column will be put in motion and pafs in flow time—viz.

    Cavalry.
    3. Companies light infantry.
    6. Battalions of the line.
    3. Companies light infantry.
    Cavalry.

*The column marches paft.*

*Fig. 178.*

No part of the column either diminifhes or increafes its relative wheeling diftances.

---

7.—When the head of the leading battalion of the column, has paffed the general one hundred yards, it may wheel to the left, and proceed diagonally towards its original ground—When the head halts at its point of formation, the rear companies of that battalion will then file to the right into line, and thus each battalion will fucceffively arrive on its ground, and each will feparately wheel up to the left into line—N. B. The whole remain in column till the head halts, the feveral battalions then break off, each towards its point in the line; and the light companies form in fecond line behind their proper battalions.

*The line is formed from column of march.*

*Fig. 178.*

8.—The

**The line prepared for the march.**

8.—The battalions will take open order.

The cavalry will march one hundred yards to the rear and form.

The light companies will move upon the right and left of the infantry—Three companies on each flank.

---

**The manual.**

9.—The manual (if thought proper for a line) each battalion taking the time from its own advanced flugel man; by one of whom, the others are regulated.

---

10.—Close order is taken.

The battalions load.

The light companies march to their original position.

The cavalry remain on the flanks of the second line, ready to support.

The right piece of cannon of each battalion moves by the rear, to the left of that battalion.

The following MOVEMENTS and MANOEUVRES comprehend the essential ones that can be required from a line, and are here arranged to the narrow compass of a review ground—They arise from the general rules already laid down, and are to be applied according to circumstances of situation, and intention. Pointing out the nature of combined operations, they are not exclusive, but only here given as one example of the many that may be deduced from the same immutable principles.

---

**The line advances, halts, and fires.**

1.—The line advances one hundred yards.

Halts.

Fires by companies.

N. B. The commander of the line, orders the cannon on the right to begin—Each battalion then fires two rounds of musketry, independent of each other. The rest of the cannon begin and end with their own battalions, and fire as fast as they can.

---

2.—The

( 77 )

| | |
|---|---|
| The line advances, halts, and fires. | 2.—The line advances one hundred and fifty yards.<br>Halts.<br>Fires grand divisions by battalions.<br>Each light company is now divided, and covers the right and left of its battalion, at fifty paces in the rear. |

---

| | |
|---|---|
| Retire by alternate half battalions of the line, and again form in one line.<br><br>Fig. 176. | 3.—The right half battalions fire in succession from the right—face about—march fifty paces lining to the right—halt—front—load—The cannon of the standing half battalion keeping up a fire, (while the others retire;) and also the right light infantry divisions, which run up and cover the interval.<br>When the right half battalions have fronted and loaded—the left fire—retire—front when in line with the others, and load. |

---

| | |
|---|---|
| Retire by alternate half battalions of the line, and again form in one line.<br><br>Fig. 176. | 4.—When the left half battalions are loaded, the right again fire, &c. &c. till the whole arrive and are formed in one line, on the first ground.<br>N. B. The cannon fire, and retire with the half battalions they belong to—Those that are advanced and whose front is clear, keep up a constant fire. The cavalry remain on the flank of the retreating half battalions. The light infantry cover the front of their proper half battalion while retreating, when it fronts and loads, they immediately return to its rear, not to interrupt its fire. |

---

| | |
|---|---|
| The three left battalions take an oblique position.<br><br>Fig. 179. | 5.—The right company of the fourth battalion wheels $\frac{1}{16}$, or any small portion of the circle to the right.<br>The other companies of the left wing, wheel each the same portion.<br>The companies of the left wing move up, and line to the right.<br>N. B. The right wing having remained on its ground, the left wing is now formed in oblique position, advanced, and facing to the right. |

---

6.—The

The right wing is thrown back to line with the left wing.

Fig. 179.

6.—The right wing goes to the right about.
    Each company wheels $\frac{1}{8}$ of the circle to the right.
    Marches to the rear.
    Halts and fronts.
    N. B. The light companies and cavalry accompany and precede thefe movements.

---

The line formed from column of march to the front.

Fig. 180.

7.—The line breaks into column of march to the left.
    The head takes a new direction by a fmall wheel to the left.
    The head, and fuch part of the column as is in the new direction, halts.
    The rear battalions of the column difengage, and march into the new direction.
    The line is formed on the original ground by wheels to the right.

---

From columns of march, the line forms part to the front, and part oblique to the rear, covering the right flank.

Fig. 181.

8.—The line marches to the front in open column of companies from the right of each battalion, diftances and lining preferved from the right.
    The light infantry cover the heads of the columns, and the cavalry follow the flank battalions.
    In the courfe of marching, the two right columns are ordered to be retired fifty and one hundred yards behind the others.
    The columns halt, and the divifions of each clofe up to their proper front ones.
    The heads of the two right columns, wheel each a fmall part of the circle to the right, and the rear of thofe columns immediately conform.
    The whole deploy on the third company of each battalion. The four left battalions in parallel line, and the two right battalions oblique to the rear, covering the right flank—The light companies continue to cover their proper battalions; and the cavalry of the right, are ftill on the right, or in its rear.

9.—The oblique battalions are attacked, and fire by subdivisions.

{ Form in two lines, oblique to the right of the rear.

Fig. 181.

Retreat of the first line through the second.

Fig. 181. }

During this, the third and fourth battalions wheel by companies 1/6 to the right, and move up into line with the oblique ones. The fifth and sixth battalions break into column of companies to the right, march and form in line behind the second and third battalions. Three light companies place themselves on each flank of the second line, and one squadron is on the flank of each line.

The battalions of the first line, fire, file firing for about two minutes, and then cease.

The first line goes to the right about, and retires; when it comes close to the second line, the companies file from their then left—pass through the second line—move on one hundred yards—halt—front—dress pivot flanks—and wheel up into line, under the protection of the second one, which is firing by platoons.

---

{ Retreat of the second line through the first.

Fig. 181. }

10.—The second line retires in the same manner that the first did; and thus repeatedly, till they are a considerable distance in the rear.

---

{ Central change of both lines.

Fig. 182. }

11.—The first line changes its front to the left, on the right company of the third battalion, so as to face its original front.

The second line makes a corresponding movement, by its right flank, in order to preserve its relative situation.

The light companies move and cover their respective battalions at fifty yards distance.

N. B. If during the above retreat of the two lines, the left is thrown back, so that the original parallel position is gained, this central movement will not become necessary.

---

12.—The

( 80 )

Second line forms in column of half battalions.

Fig. 183.

12.—The two battalions of the second line march to the right, cover, but out-flank the two right battalions of the first line, and then form in close column of half battalions, behind their respective left companies. The right one of these columns, moves up to the rear of the grenadiers of the first battalion within sixty paces.

---

Attack in echellon from the right.

Fig. 183.

13.—The whole move on fifty yards, and halt; the cannon begin to fire, and the three columns of the second line deploy. The right half battalion of the first line, and the column behind it move on—The other three half battalions of the right wing follow it successively in echellon, at sixty paces from each other, and then the left wing which remains in line—The second line supports in corresponding echellons, so that its right one will be about in the prolongation of the left wing.

The whole halt; the light companies are behind their respective battalions; the cavalry behind the flanks; and the half battalion in column deploys to the right, and out-flanks the advanced echellon.

The cannon and right echellon fire two rounds by platoons, and the file firing.

---

The right flank retired.

Fig. 183.

14.—The two left companies of the first line are thrown back on their left, so as to give a position considerably oblique to the rear—The center echellon of the second line stands fast, the other two, immediately join it—The out-flanking half battalion and cavalry stand fast; the advanced echellons all retire, pass through the second line, which is fronted and halted to protect the movement, and form in line in the oblique position given from the left as the point d'appui, in this manner, throwing back the right of the whole. When the retiring echellons begin to pass through the second line, the whole left wing, except the two flank companies immediately faces about, and takes up the oblique position. The out-flanking half battalion fires when the retiring echellons arrive near it, follows them at the distance of one hundred paces, passes the second line, and places itself in column behind the right flank.

15.—The

( 81 )

The lines change position on the center.
> 15.—The firſt line reſumes its original parallel front by the right moving up, and the left falling back on the center of the whole. The ſecond line takes its relative poſition—Both lines if neceſſary, take ground to the right in columns, and again form up.

---

Attack in line.
Fig. 184.
> 16.—The whole advance—The right wing and ſecond line in front; and the left wing follows the right one in echellon of half battalions.
> The whole halt—The right wing fires by half battalions, and that in column deploys to the right.
> File firing—during which,
> The left battalions move up into line.
> Advance in line—charge bayonets—halt—light infantry purſue towards the center—cavalry charge in front of the flank.

---

> 17.—Half battalions of the ſecond line, march and cover the ſecond and third battalions—The cavalry return and form on the flanks of the firſt line—Three light companies form on each flank of the ſecond line.

---

The whole advance—halt—take open order—ſalute.

---

## Idea of MANOEUVRE for a small Corps, which marches to attack an Enemy posted, and retires after being repulsed in the Attack.

A Corps posted at C, has an advanced party of light infantry in a hollow at A, from whence they first fire on the approaching enemy. They then retire to B; and in proportion as the enemy moves on, they fall back, skirmishing with the flankers, but never crossing the hollow till they have joined their corps at C.

The column of five battalions, and one regiment of cavalry is assembled at D: an advanced guard is formed of the grenadier companies—The vanguard moves from the left, forms in line when across the hollow, is followed at the distance of three hundred yards, by the column which has marched from the right by companies; the column halts at E, and the vanguard at e, on receiving a fire from the enemy at A, who immediately falls back.

Three companies from the front of the column are ordered up to replace at e, the vanguard, which is detached to f to take post behind the Ravine: when the passage is thus insured, the cavalry move quickly by the high road to F, and there form in line: the vanguard cannonades the enemy at B, and the column marches by half companies to F, where it forms in line G by extending to the left. The companies which were posted in e, follow the rear, and join on the right; and when the column is forming into line G, the cavalry advance and cover its left flank; and the vanguard which has continued skirmishing, also move to their left, so as to arrive at g when the line is formed.

The

( 83 )

The line now breaks to the left by companies; the advanced guard marches to h, and the column to H, where it forms in clofe column, and fees the enemy pofted at C.

The column being formed with the left in front, and halted on the right of its ground, deploys into the oblique line H, on the right company of each battalion, which company is now in the rear of each.

The line advances to I, halts for an inftant; and the vanguard fkirting the wood moves on to i.

Each battalion wheels to its right, a fpace fufficient to bring it parallel to the general front of the enemy. The line advances in echellon, led by the fecond battalion from the left; the reft keep retired fifty or fixty paces behind each other. They move on brifkly to the attack; the leading one when at a proper diftance, begins to fire by half battalion, and the others (except the left one, which halts when the leading one halts, and does not fire) fucceffively as they come up in line with it, alfo engage at L. The van edges away to l, but does not fire; and the cannon which are with it, as well as thofe on the left, do fire in an oblique direction.

---

The attack not fucceeding, it becomes neceffary to retreat.

The fecond battalion from the left retires; and thus the others fucceffively, till the whole are in echellon, at fifty yards diftance from each other, except the left battalion, which remains one hundred yards from its adjoining one. The right battalion, as foon as it has faced about and retires, wheels gradually to its left, and gains the fkirt of the wood, where it fronts; the other battalions alfo, by gradually wheeling up to the left, Plate 25. in the courfe of their retreat, gain the pofition M, and front, being protected in this operation by the vanguard placed in m, by the cannon, and by the cavalry in L, through which part of the infantry do march. The cavalry now place themfelves on the left, and cannon are with each wing.

From the pofition M, the two battalions of the right are ordered to retire by files from the right of companies, through the wood to the open

ground

ground at N. The three left battalions retire in line; the grenadier companies cover the right flank, and the cavalry the left.

From N, the corps retreats to O, and the cannon are posted at O.

From O, the cavalry retire briskly along the road; the three left battalions follow them, and are protected in this operation by the two right ones, who do not fall back, till the enemy in course of following the left wing, cannonade them from near S.

The whole corps now assembles at P, and the grenadiers preserve their post at p.

The corps takes the position R, by marching from the right of lines.

The relative movements of the cavalry are marked at *L, M, N, O, P, R*; and those of the van and rear guards are also marked, e, f, g, h, i, l, m, n, o, p, r.

A detachment x, guards the narrow pass, and advances and retires with the corps.

When the retreat is making, the enemy detach and follow the grenadiers and the column, the rear of which they will cannonade from the height S, where the pursuit naturally ends.

The marches and formations of the battalions, should be made without hurry, at a steady uniform pace, such as is proper for a considerable body of infantry.

The movements of the cavalry should be made with precision, firmness and due celerity.

The ground nearly represents part of the Phœnix Park, Dublin—The manœuvre was executed by the garrison, September 1787. It is meant to show the application of regular changes of situation in broken ground; and is given as an example of exercise that may be varied, ad infinitum.

The precision and exactness required in the Movements laid down in the preceding Work, and in this Appendix are attainable and cannot be unnecessary to the British troops: for they reconcile order and exertion, which is the great object of all military arrangement. Where they compleatly apply to ground and circumstances, their justness and superiority will always decide: where they do not fully apply, the less they are departed from, the nearer to perfection will the body that they direct attain; and there is hardly any situation, in which the single battalion, its subdivisions, or even a proportion of the line, cannot be regulated by them.

<small>General conclusion.</small>

There are unquestionably many occasions, where it may be improper rigidly to follow all the formal rules of method; but it would be a bad general system which was not built on order, and had irregularity for its foundation. Quick movements of small bodies, may when necessary be demanded, and the troops should be accordingly prepared; but great care is to be taken against throwing any considerable body of infantry into disorder by hurrying them: in the presence of the smallest corps of determined cavalry, it will always be fatal. Where a rabble, or a timid and inferior enemy is to be attacked, and can be got at, time often would be lost in acting with too much regularity; spirit and activity are then generally sufficient: but if steady troops are to be encountered, it is on firmness and combined exertion that we must depend for success.

---

Should we find peculiar difficulties in the correct execution of what is here required; yet even a near approximation and application of these principles, would soon produce most evident advantage.

If the divisions of a column, marching in the alignement, are not in a situation to halt at their just distances, all at the same instant: yet they may and ought truly to do so in succession, by word from their several officers.

Should the battalions marching in line be not so exactly dressed, as to halt correctly all at the same instant: yet they may successively be ordered so to do.

Should the pivots of a column marching in the alignement not exactly cover: yet the mounted officers can, and must dress them truly, before the formation of the line is attempted.

<small>General principles.</small>

The

The indispensable rule of the staff officers being mounted, should be revived and enforced, and the adjutants enabled to comply with such rules.

The internal formation of our battalions, should be deduced from principles of service, and not of parade.

Our open order should be abolished; our files should be closed; and the old ideas of firmness, compactness, and mutual support, be restored and held sacred.

The ordinary and quick steps of march, each uniform in its length and cadence, should from habit be acquired.

The necessity of all bodies whatever, great or small, making their marches and formations in determined lines, should be inforced; and the method of taking those lines with exactness, should be ascertained.

The importance of bodies being able to march for some distance in file, without opening out, should be understood and practised; and also the necessity of inclining in a parallel line, without altering the front of the battalion.

<sub>General principles.</sub> Officers should be convinced of the propriety, and steadied in the practice of pivot covering, and of preserving their true distances; of commands being circulated and executed quick, decided, and with precision; of the justness of movements and formations of the battalion, on given and determined objects, from the influence they have on those of the line—They should avoid hurry, but aim at quickness consistent with order.

Our general movements and modes of execution, should be such as are applicable to the line, and those of the battalion reconciled to, and deduced from them.

Music, drum, or fife, should not be used or trusted to in the instruction of the recruit, officer, or battalion; nor in manœuvre; on occasions of parade they may be permitted.

Movements and alterations of situation should be divided into their distinct and separate parts; so that one operation may be fully compleated and paused upon, before the commencement of another.

Fixed points should be clearly ascertained before bodies march or form upon them.

Close order should always be considered as the essential and fundamental one. Open order is only accidental, and for parade.

The

The foldier muft not be puzzled with various and unprofitable ways of attaining the fame end.

The fame principles fhould extend equally to the movements of cavalry, as of infantry.

---

Such among many others, are circumftances that cannot be difpenfed with, and which we ought to follow and eftablifh; their oppofites in many inftances prevail in our fervice, and fhould be difufed and forbid. But if fuch obfervances are effential to the movements of the fingle battalion, how much more important and neceffary are they to the manœuvres of more confiderable bodies of troops; and how confpicuous would this appear if we had the fame opportunities of affembly, which all well difciplined armies annually have! for it is by fuch means only, that in time of peace the foldier is at leifure formed; that talents can be known and brought forth, and that a judgment can be made of the future abilities of the officer, when war calls him into real fervice. On this advantage to the wifhed for degree, we can hardly calculate; for in moft of our foreign ftations, in England, and in Scotland, the infantry are fo few, or fo difperfed, that no means of collecting them can be expected; and it is only in the capital, Gibraltar, and in Dublin, where with any convenience numbers are affembled. *Neceffity of affembling large bodies.*

In the CAPITAL, there feems no military reafon why the guards, permanently quartered, and fufficiently numerous, fhould not be trained on one uniform fyftem; and become from their fituation, a model for the reft of the army. *London.*

In GIBRALTAR, however well calculated for forming troops to many effential parts of fervice; yet the very confined ftate of the garrifon, will allow them only to practife the firft rudiments of movement; and prevents them from feeing the extent and confequence of confiderable manœuvres. *Gibraltar.*

In IRELAND, they are better fituated; for befides the ftations of Cork, Kinfale, and Galway; DUBLIN particularly offers every opportunity of forming troops, and of fhowing the neceffity of general fyftem.—The ground for exercife is varied and extenfive, the garrifon is confiderable, for two months in each fpring it might without much expence be increafed, and avoiding all matters of parade, or of the drill, that period might be dedicated to give ideas of manœuvre and the movements of great bodies. In *Dublin.*

the autumn also the same exercises might again take place; and although formed battalions only should be brought to this quarter, yet the intervening periods would give sufficient time for whatever detailed drilling particular regiments might require—Was such a mode adopted, and followed out, it must be attended with the most salutary effects in the course of a few years, as the greatest part of the infantry of the army would circulate during that time through Dublin.

---

**Previous system necessary.**

But to enable us to proceed with success to so desirable an object, it is previously necessary, that one general and detailed system should be established, derived it is hoped from that which experience and success has shown to be the best; and when such is by authority prescribed, it should be implicitly followed, rigidly maintained, and no regimental deviation from it in the smallest instance permitted.

**Consequent advantages.**

Were we once steadied in true principles, there can be no doubt that we should arrive at the fullest proficiency, for an immutable line of conduct being prescribed and followed—caprice and change would no longer exist—Uncertainty which is so harrassing to all concerned, would cease—General uniformity would take place—Much laudable zeal and ability would be directed aright—Officers no longer acting on the temporary ideas of individuals would soon perceive the first stages of the profession, to be those of science, and that without their correct aid, the combinations of the general cannot be executed—The just execution of such movements as can be required, or are really applicable to the service in the field, would be unalterably established—And explicit regulation instead of tending to confine or to cramp genius, would give fuller scope to military talents in the application of true principles to the boundless variety which time, numbers, situation of ground, and other circumstances will always present.

F I N I S.

# APPENDIX.

## CONTENTS.

General CIRCUMSTANCES, WORDS of COMMAND, and PARTICULARS of EXECUTION, necessary in the following MOVEMENTS.

### COMPANY.

| | Page |
|---|---|
| WHEN to march, halt, dress | 10 |
| break to a flank, wheel up to form, march in file | 11 |
| halt, and front, countermarch, incline | 12 |

### BATTALION or LINE.

| | Page |
|---|---|
| When to march | 13 |
| halt, file | 14 |
| incline, change direction | 15 |
| retire, front, and dress, retire by files from the right of platoons | 16 |
| Pass an obstacle | 17 |
| Move up after passing the obstacle | 18 |
| Passage of battalions or lines | 19 |

### OPEN COLUMN.

| | Page |
|---|---|
| When the battalion breaks to either flank, To march from a flank to the {front, rear} | 20 |
| To form column on a central division, When the column is to {march, halt, form in line} | 21 |
| When the divisions of the column successively wheel | 22 |
| Wheel of unequal divisions in column | 23 |
| When the head halts, and the rear files into line | 24 |
| When the rear battalions of a column march into line | 25 |
| Change of front by countermarch of divisions | 27 |
| Forming to the rear on the {front division, rear division} | ib. |
| When the column files to the flank to change position | 28 |
| When a considerable column changes position | 30 |

|  | Page |
|---|---|
| To form an oblique line on a central division | 31 |
| To form an oblique line on the front division of the column | 36 |
| To form an oblique line on the rear division of the column | ib. |
| To form the square | 38 |
| From the square to form an oblong | ib. |
| From battalion in line to form an oblong | 39 |
| From the square or oblong to form in line | ib. |
| When the column of march makes front to both flanks | ib. |
| When the oblong is composed of several battalions | 40 |
| When the sides of the oblong are six deep | ib. |
| When the oblong forms in line | ib. |
| Change of position by oblique marching | ib. |

### ECHELLON.

|  | Page |
|---|---|
| When the battalion is marching, and the right wing is to form echellon to the rear | 41 |
| When halted, and that the echellon is ordered to form oblique line to the right | ib. |
| When the oblique line is to resume echellon | ib. |
| When the echellon halted, and formed to the rear—is ordered to advance, and form echellon to the front on the same division | 42 |

### CLOSE COLUMN.

|  | Page |
|---|---|
| From line to form close column { in front of right division / left division | 43 |
| { in rear of right division / left division | 44 |
| on a central division } |  |
| When the line forms several close columns | 45 |
| From column of march to form close column | 45 |
| From battalion column to form line on a { front division / rear division / central division } | 46 |
| The close column halted to take a new direction | 49 |
| When the column countermarches by divisions | ib. |
| When a considerable column of march, forms close column, and deploys into line | 50 |
| Formation in line from several columns | 52 |
| Taking distances from the { rear / front / central } division of the column | 53 |
| When the column forms line in the prolongation of either flank | 54 |
| When the battalion forms double column from the center | ib. |
| Double column of several battalions | 55 |
| When the double column extends into line | ib. |
| When the battalions of second line, form double column on the center companies | ib. |
| When the alternate companies of the second line double | ib. |
| When the line retreats in several squares or oblongs | 56 |
| When to retreat part in line, part in column | ib. |
| When the solid column makes front on all sides | ib. |

| | Page |
|---|---|
| Form of review and movements necessary for the practice of a single battalion, and of a small corps | 57 |
| Idea of a manœuvre for a small corps | 82 |
| Conclusion | 85 |

# ERRATA.

| Page | Line | |
|---|---|---|
| 21 | 26 | After—Commander of—insert—*the*. |
| 43 | 2 | on margin, instead of Defilé, read *File*. |
| 60 | 8 | instead of company—read—*companies*. |
| 70 | 4 | instead of composed—read—*arranged*. |
| 77 | 10 | instead of step—read—*stop*. |
| 82 | 15 | instead of broke—read—*broken*. |
| 85 | 23 | instead of alignemt—read *alignement*. |
| 122 | 23 | instead of squadron—read—*squadrons*. |
| 142 | 26 | after but—insert *if*. |
| 160 | 13 | after original position, insert—O. |
| 221 | 4 | instead of where—read *were*. |
| 242 | 30 | instead of 5th, read *1st*. |
| 255 | 22 | instead of briskly, read *rapidly*. |
| 260 | 8 | instead of Guningen, read *Gruningen*. |

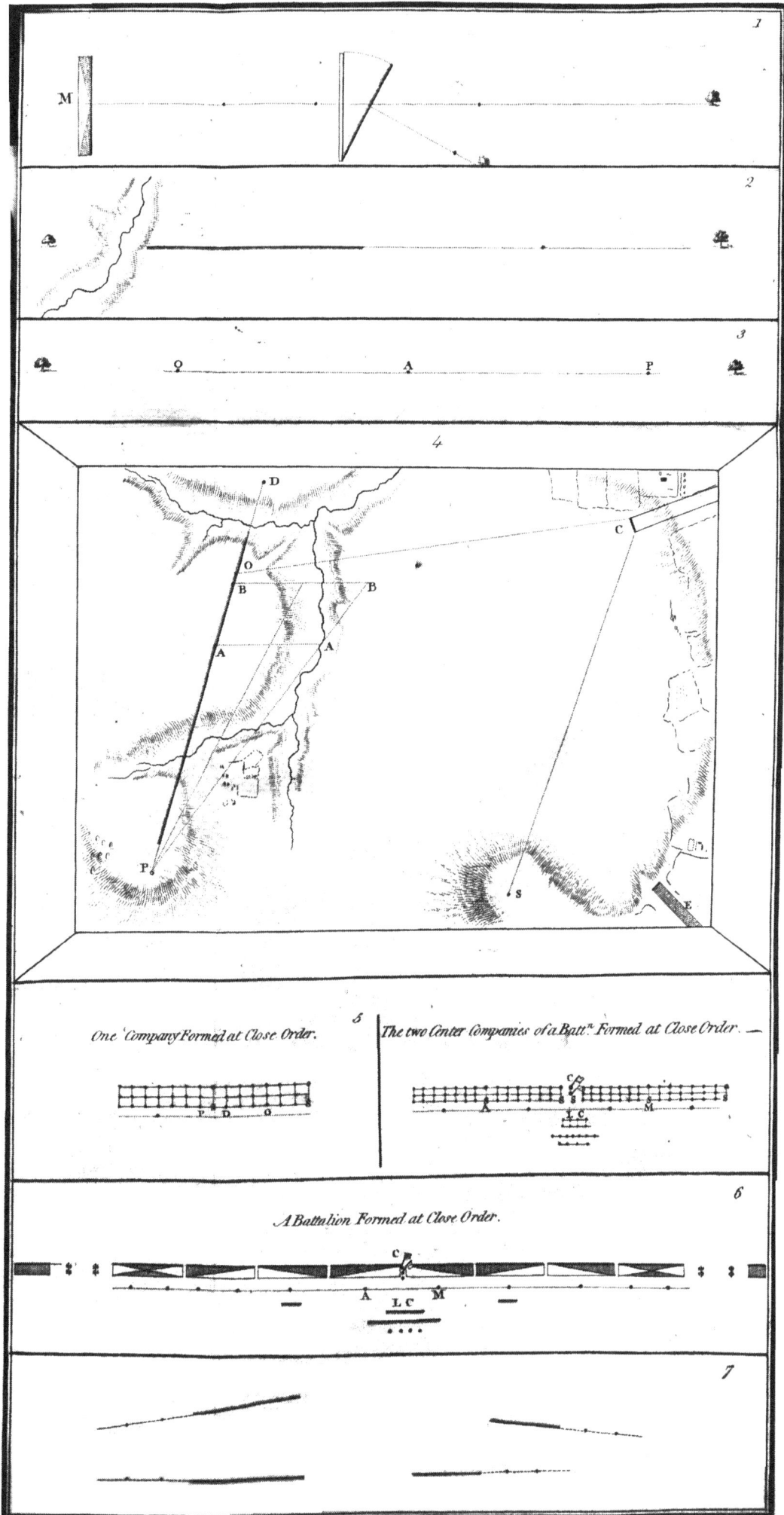

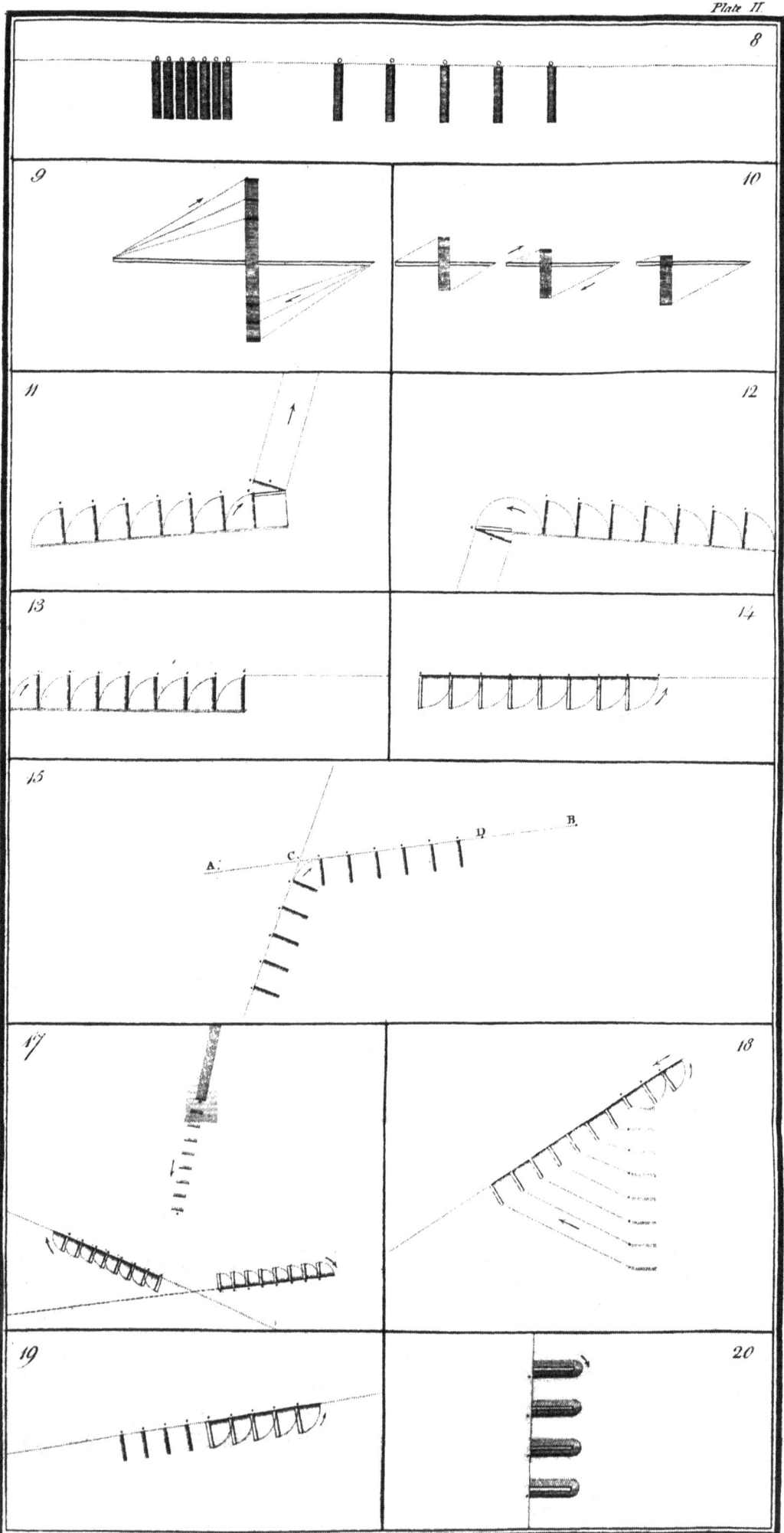

Plate II.

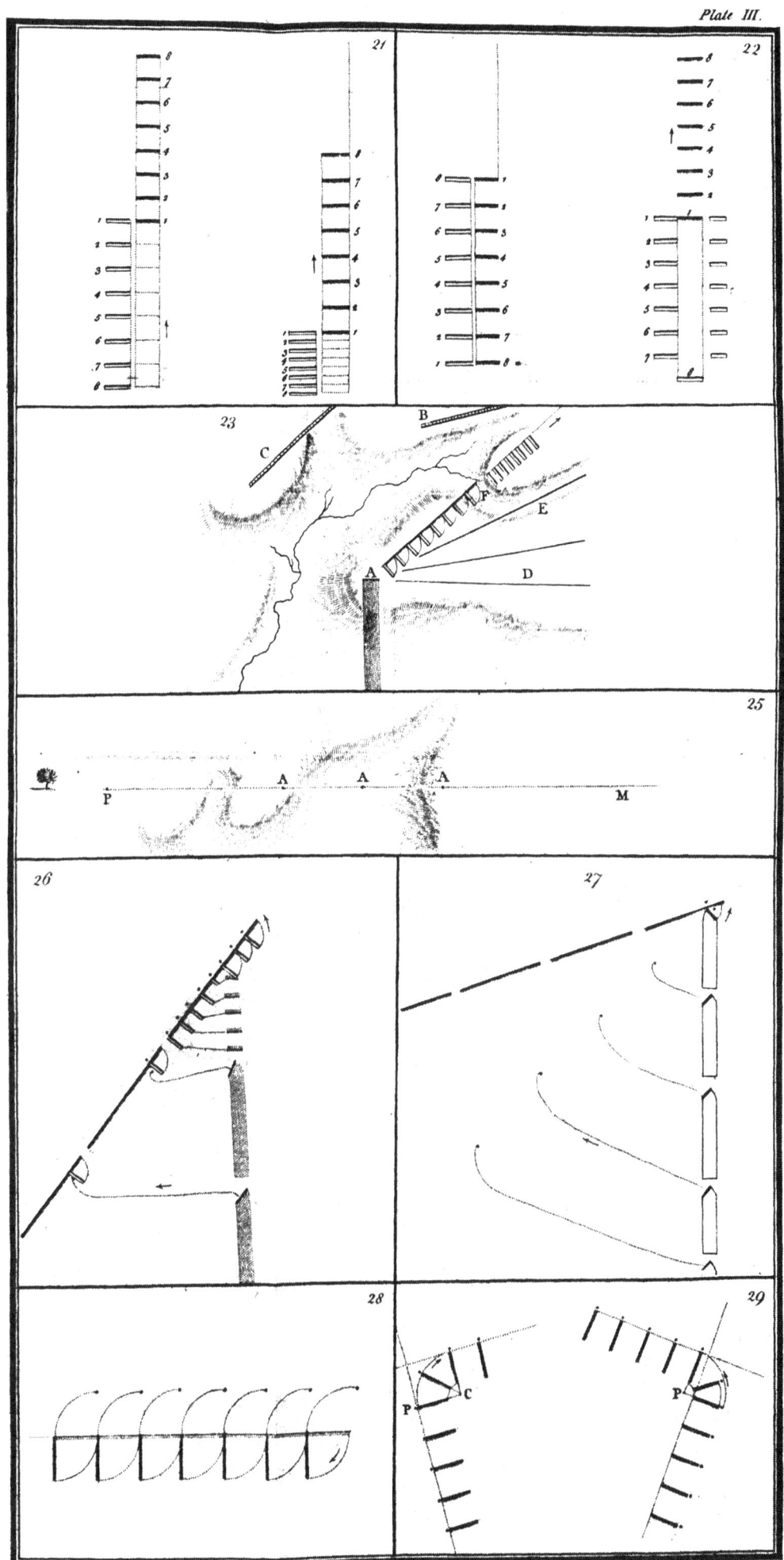

Plate III.

Plate IV.

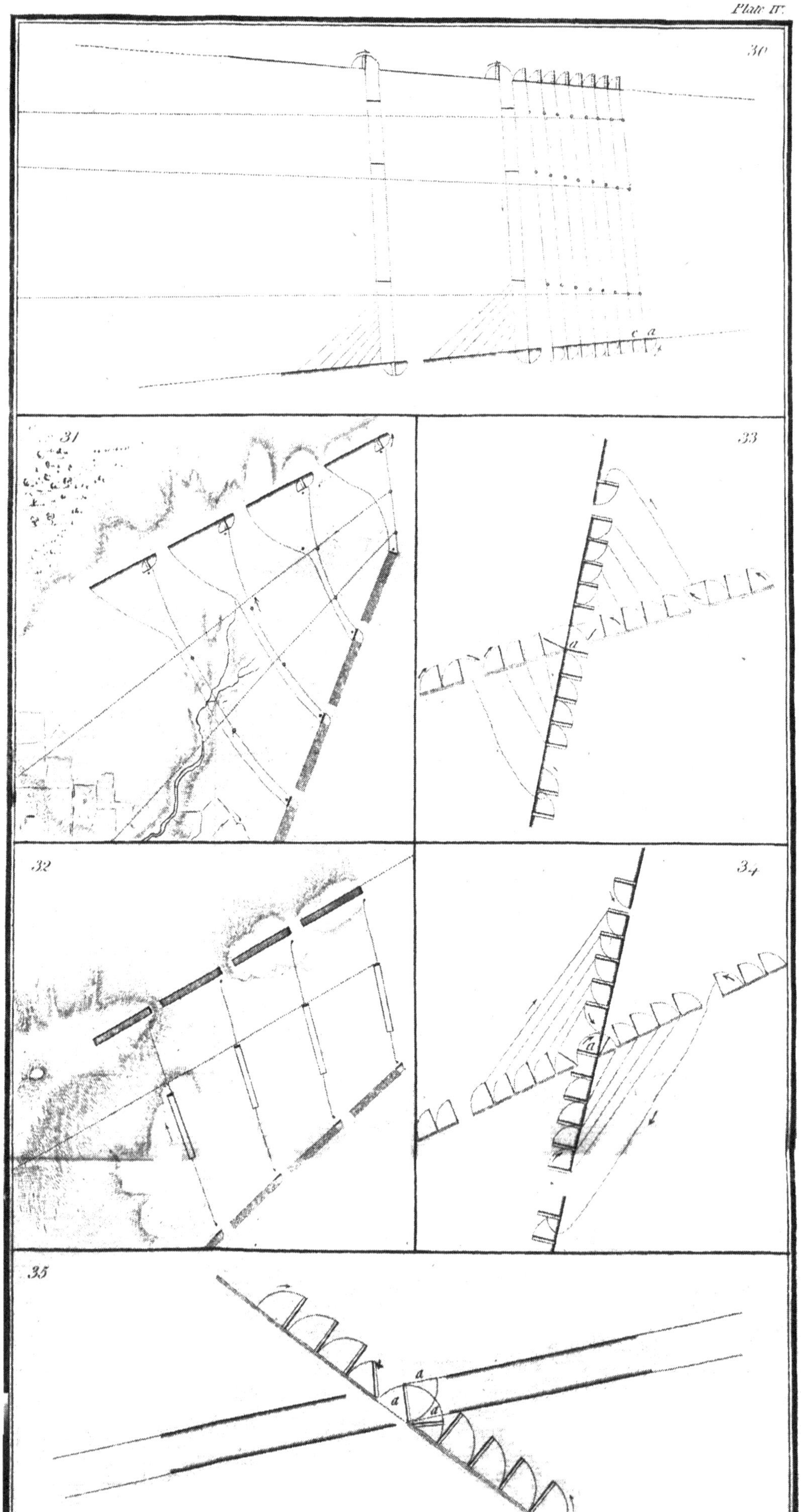

30

31

33

32

34

35

Plate V.

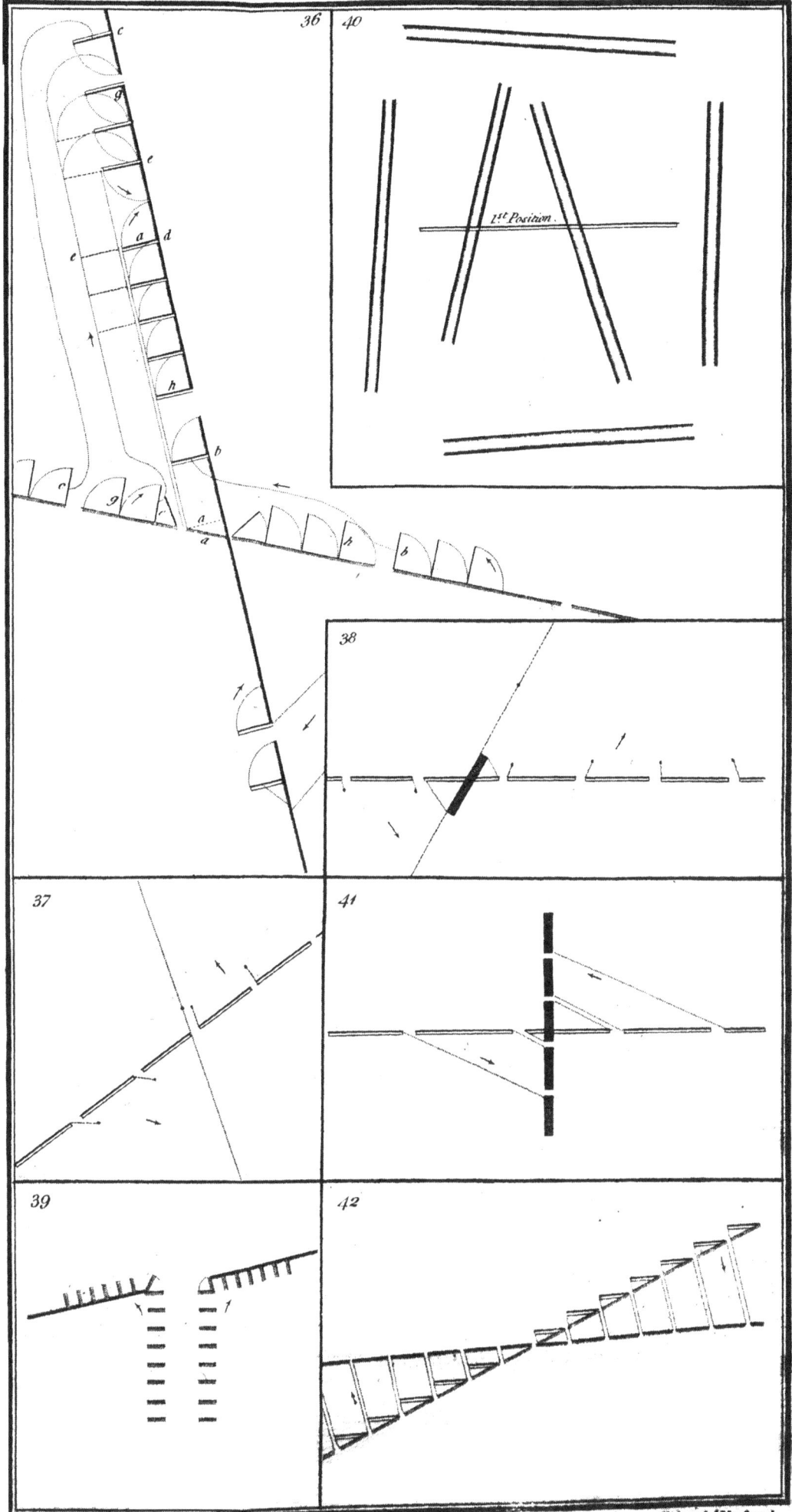

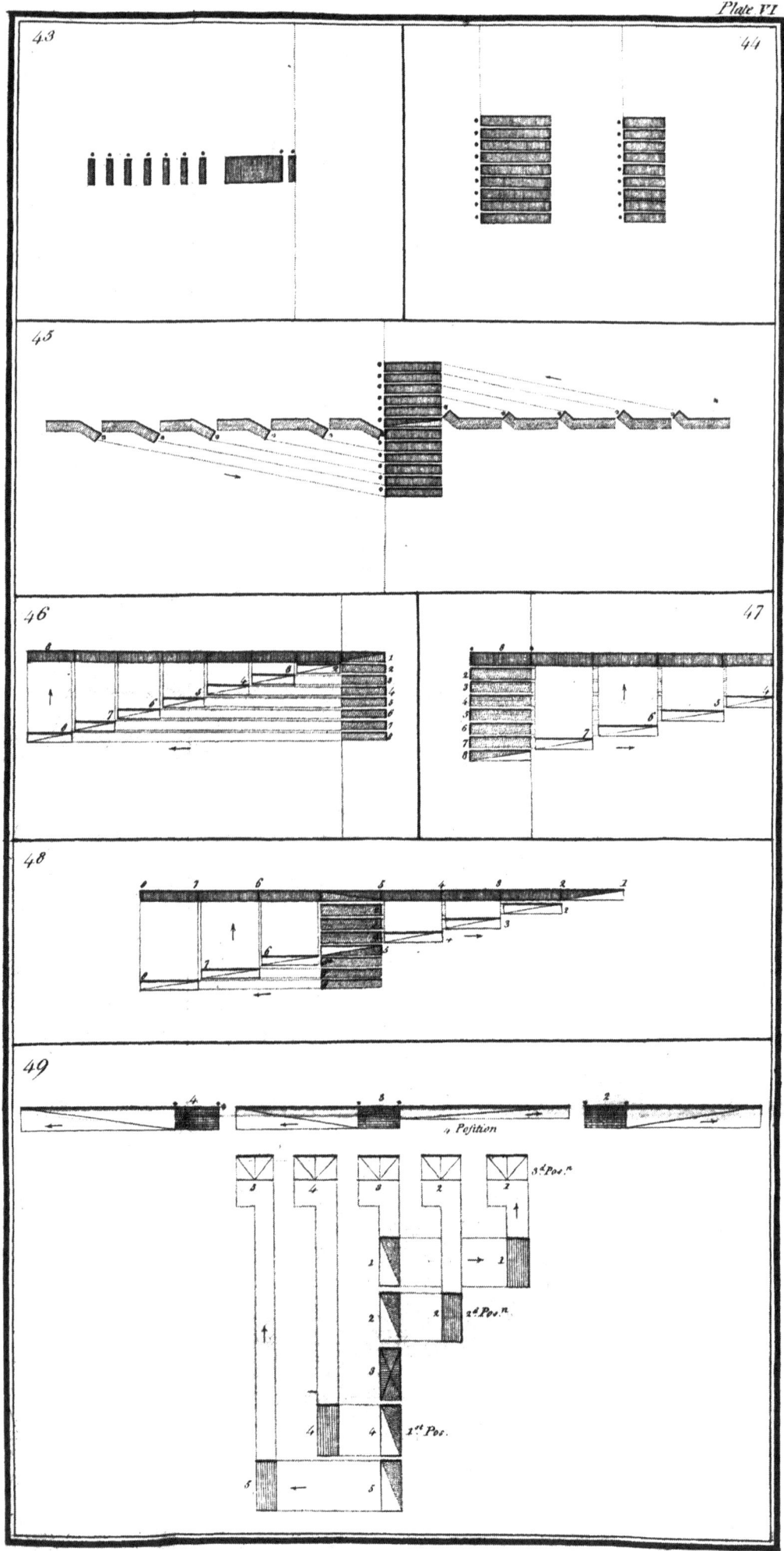

Plate VII.

*True Line of Formation.*

*False Lines of Formation.*

J. Neele sculpt. Strand.

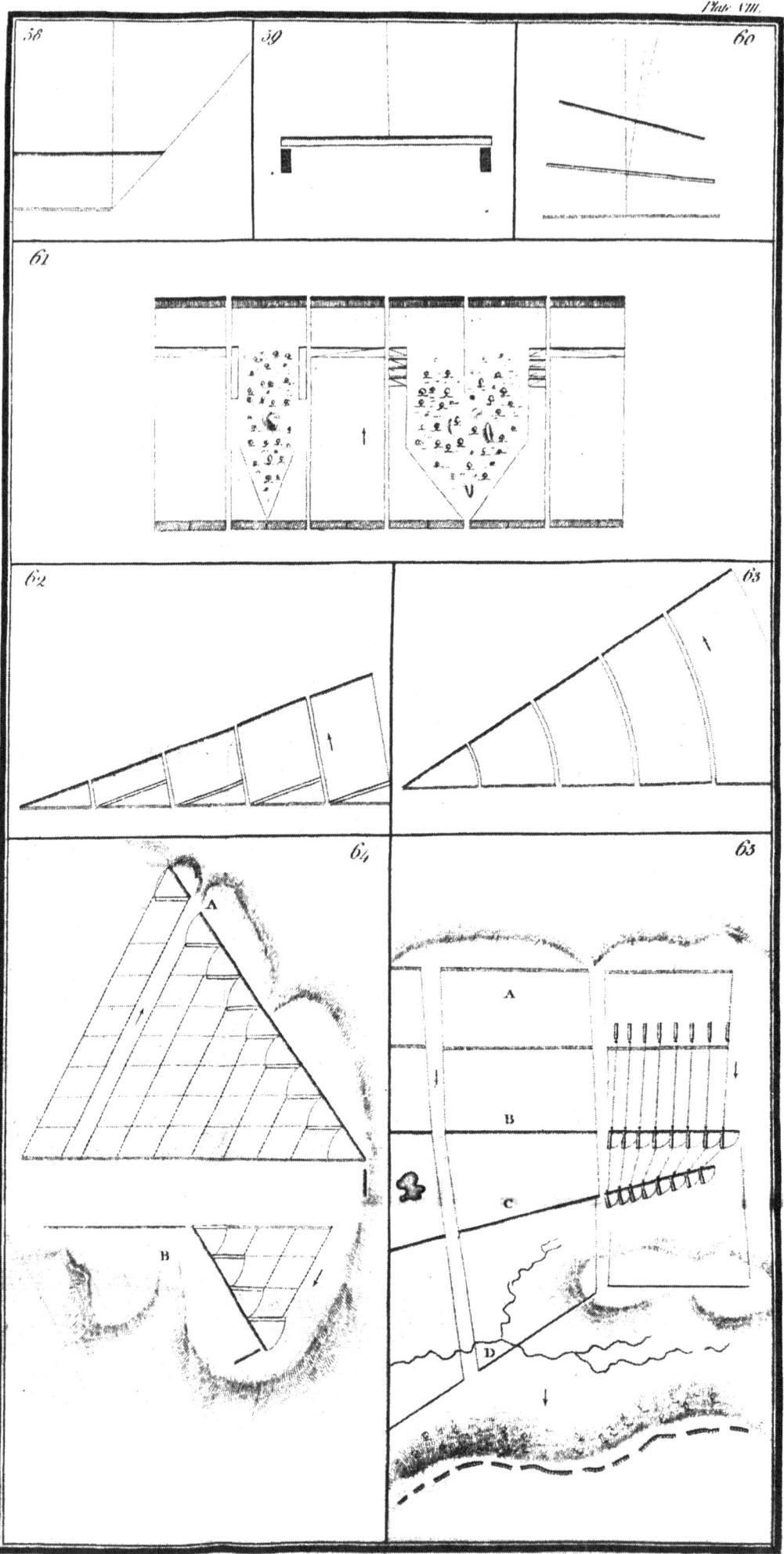

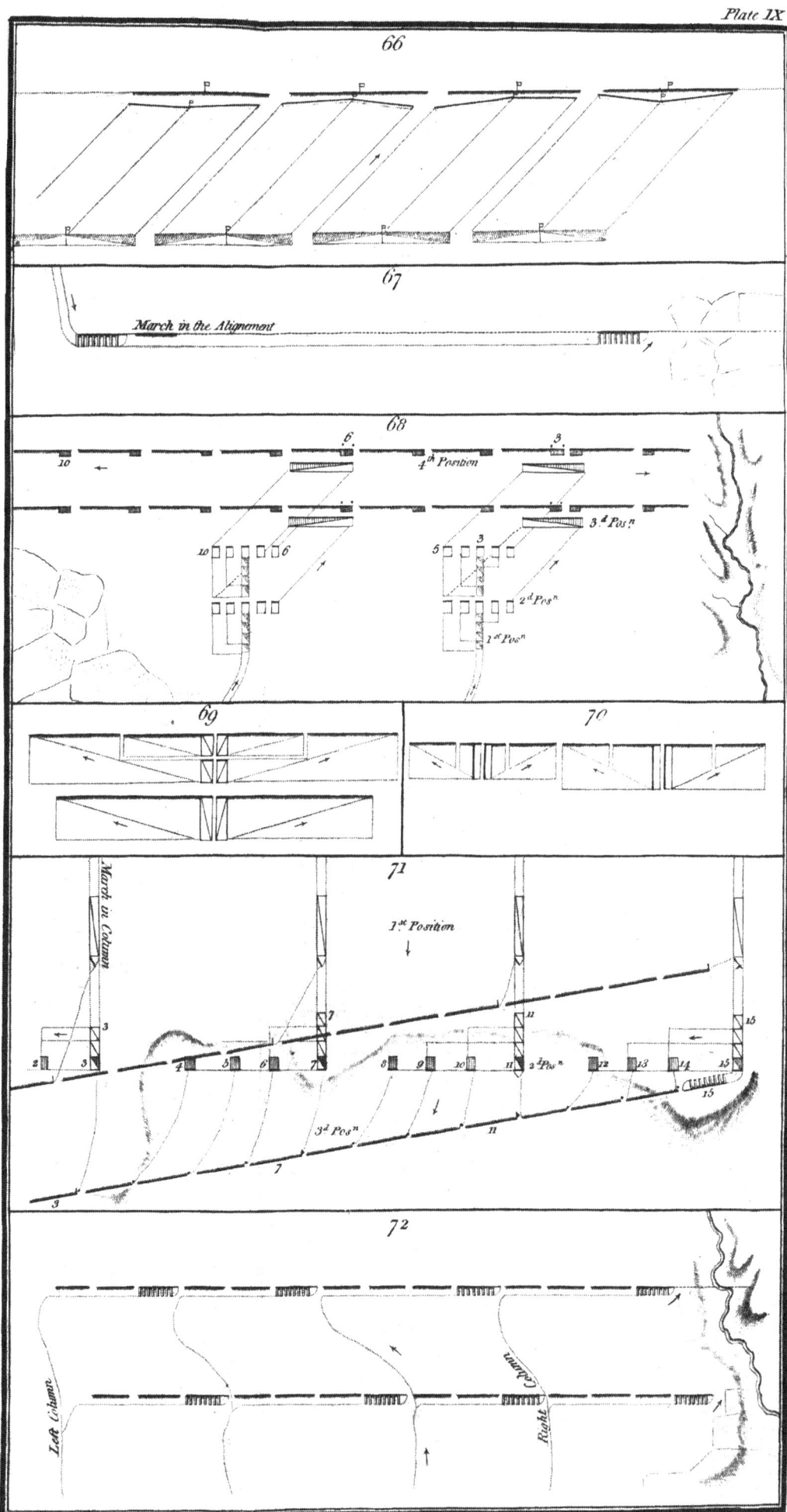

Plate X.

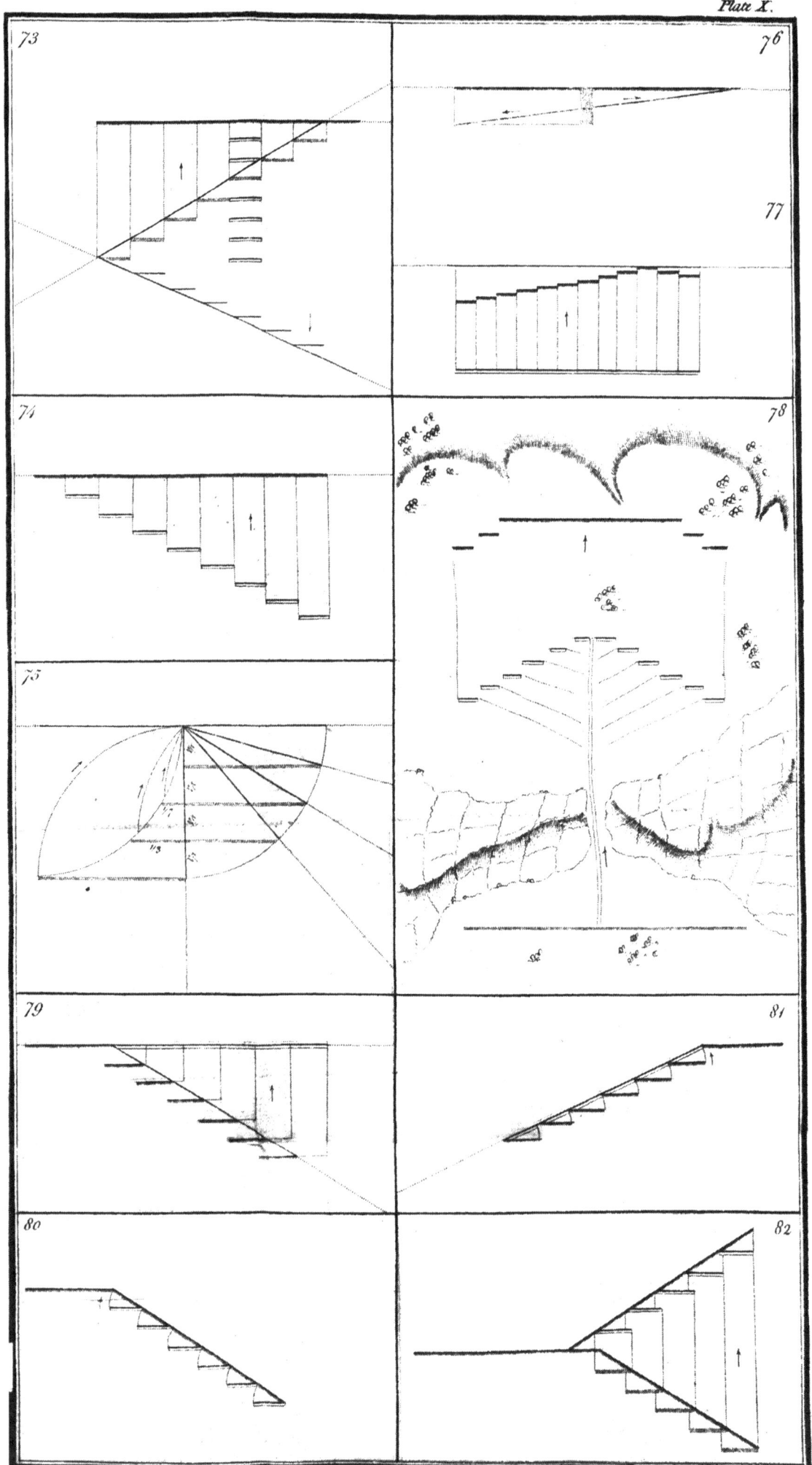

Plate XI.

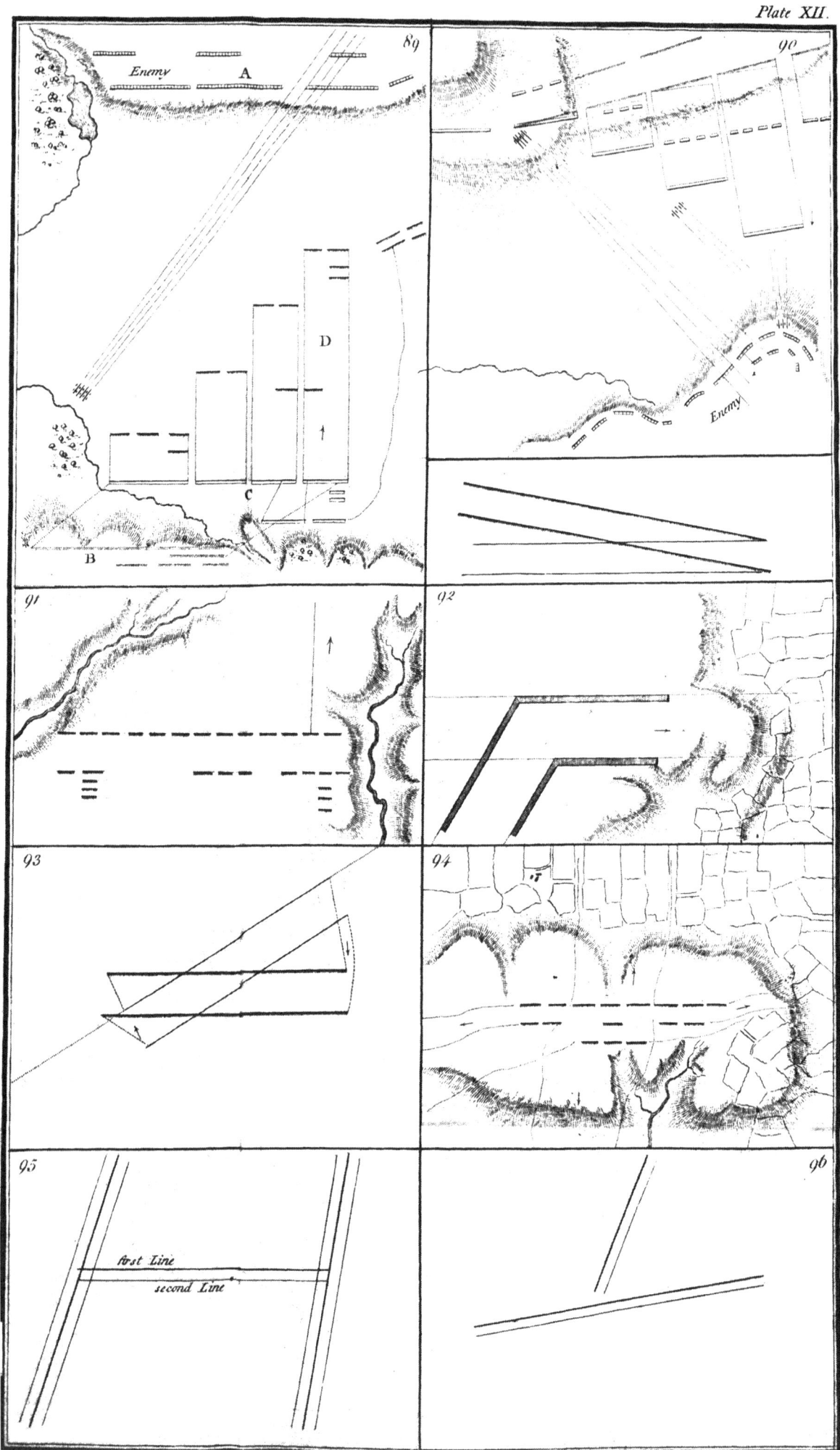

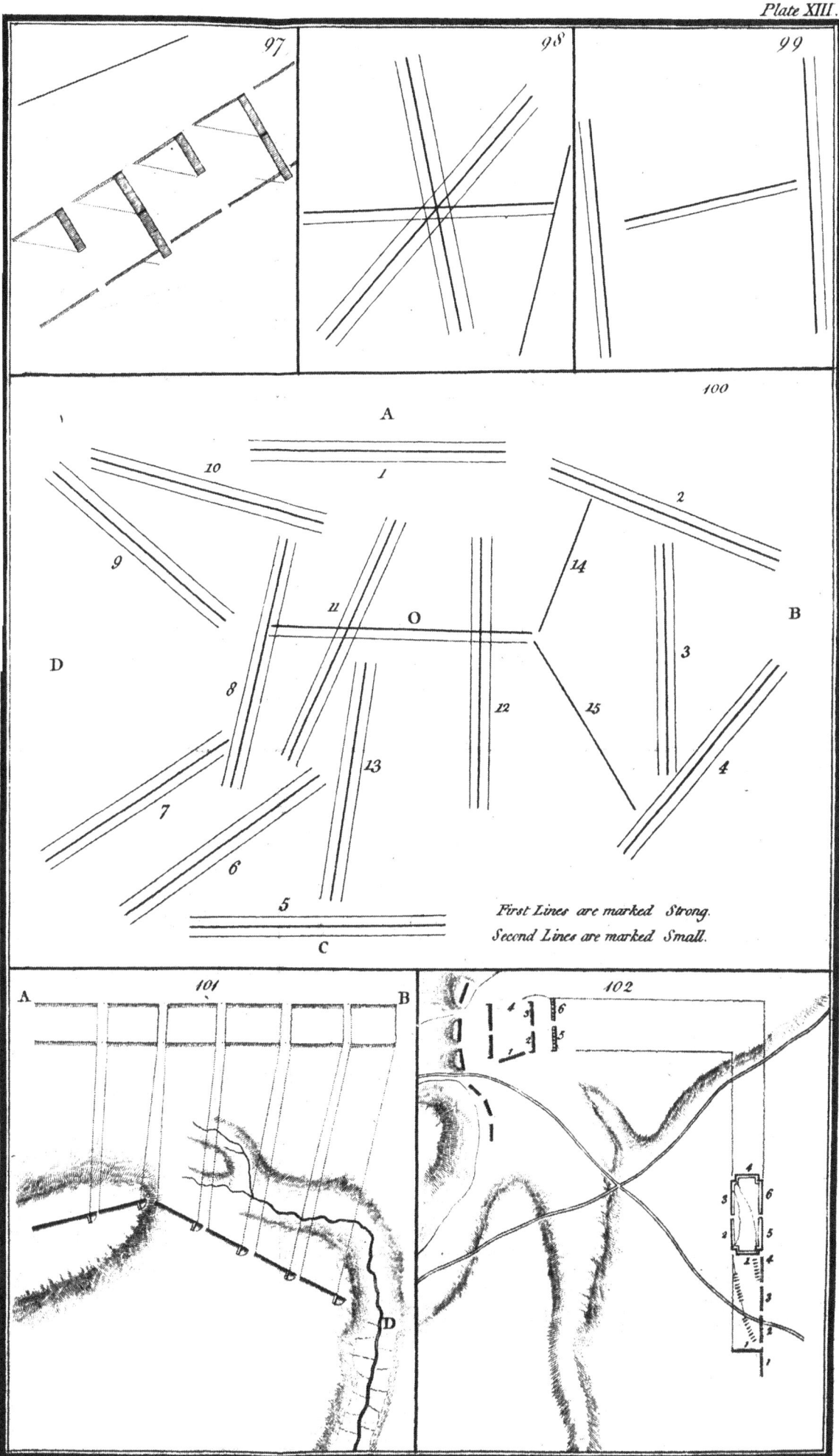

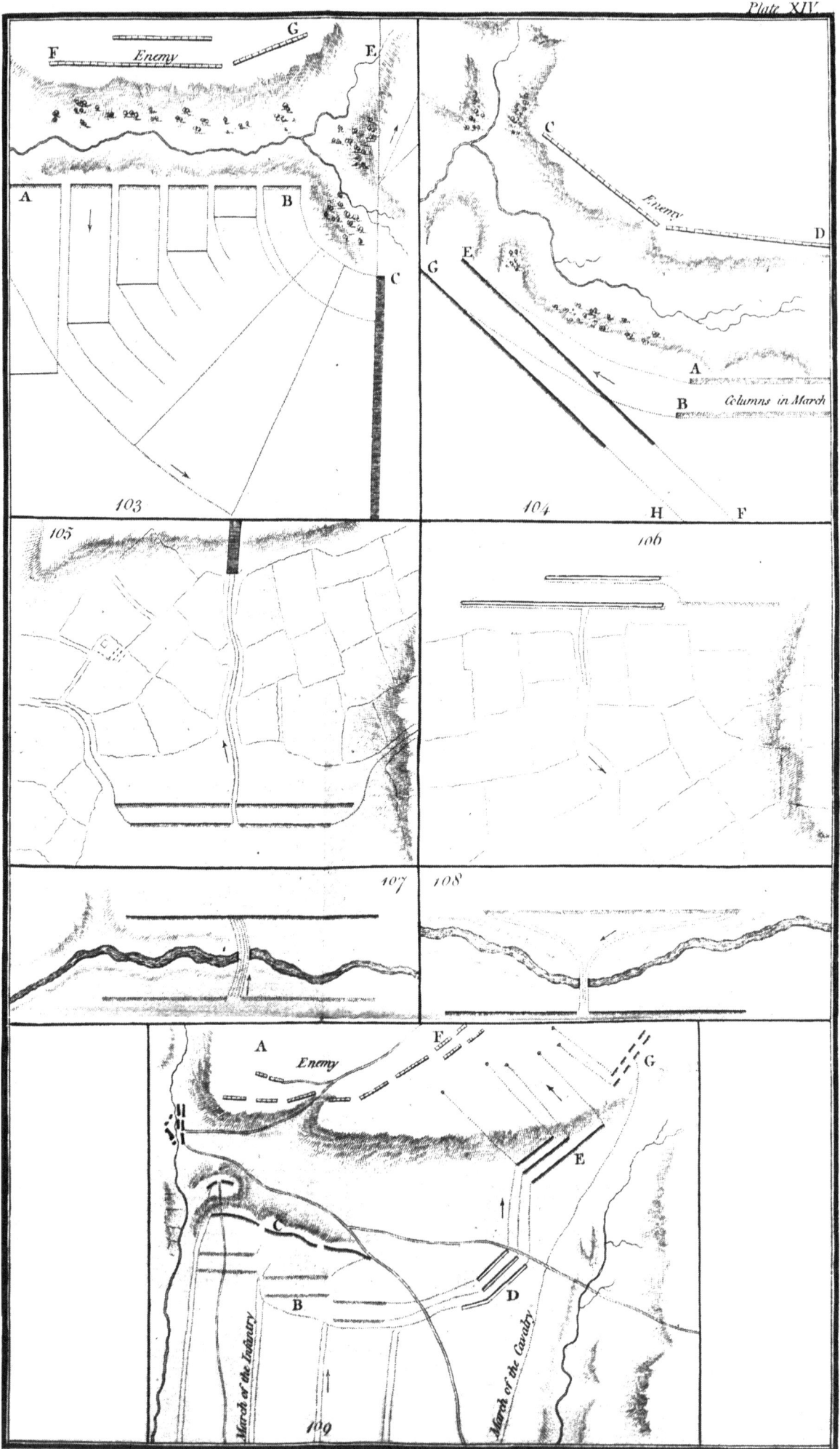

Plate XV.

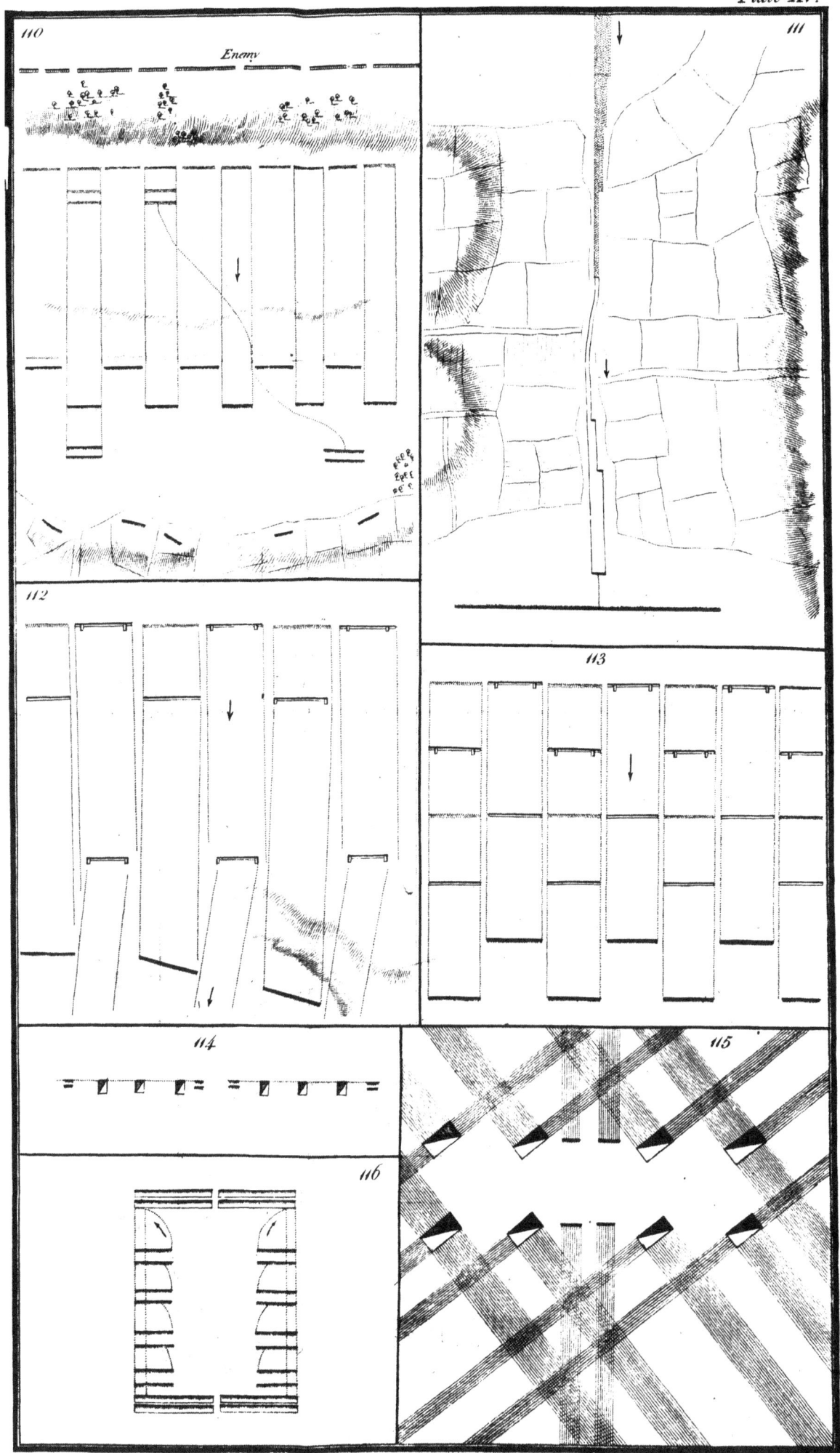

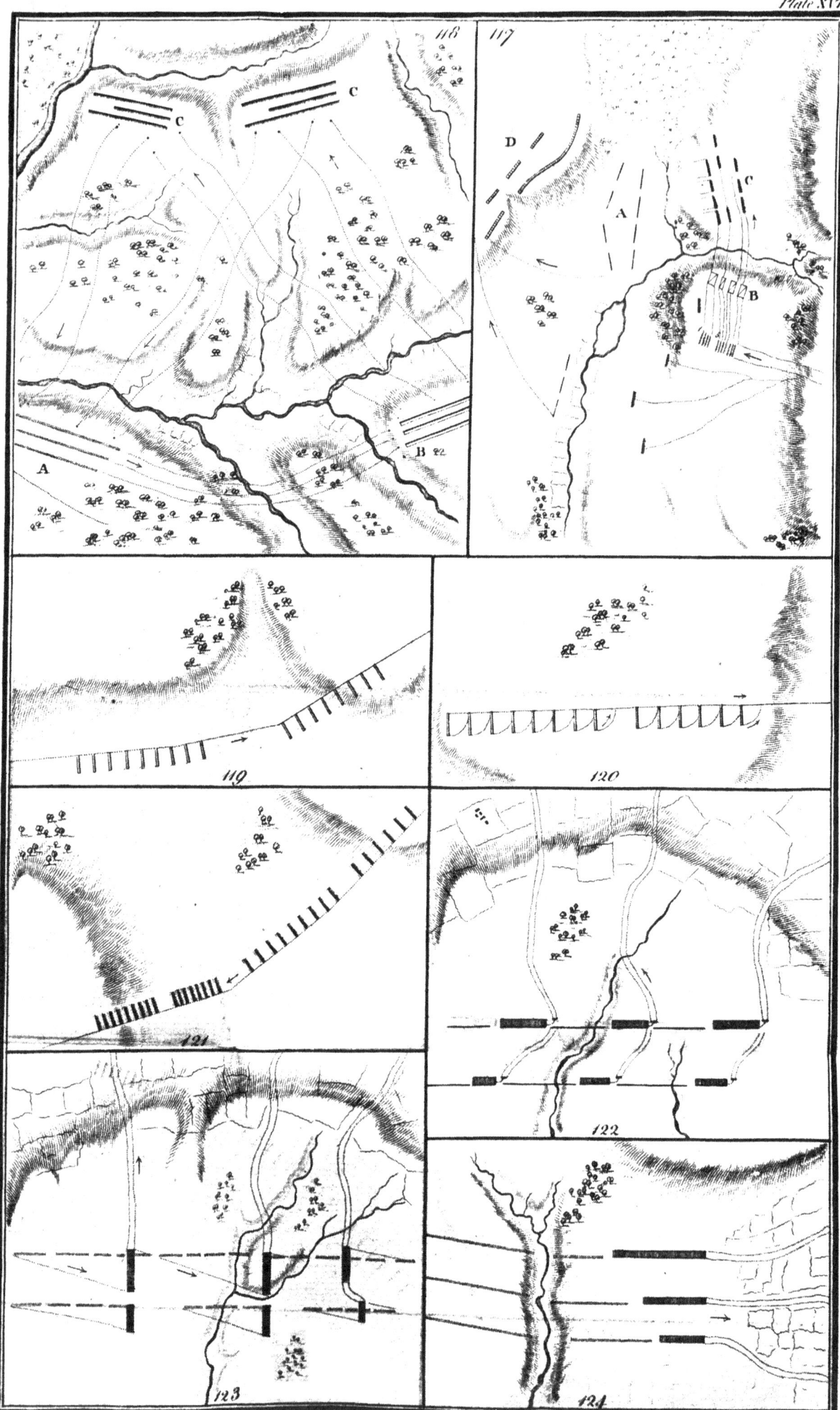

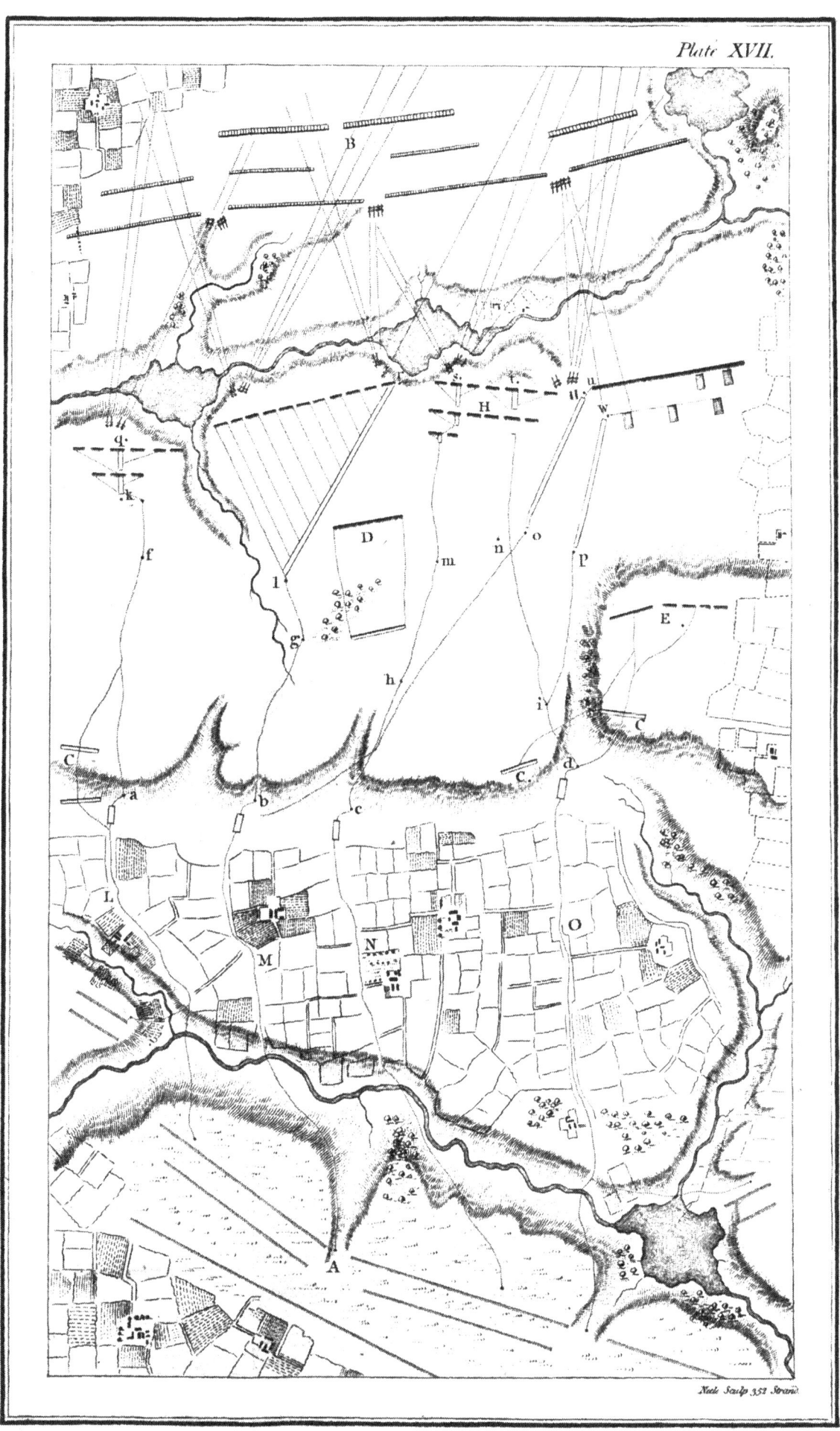

Plate XVII.

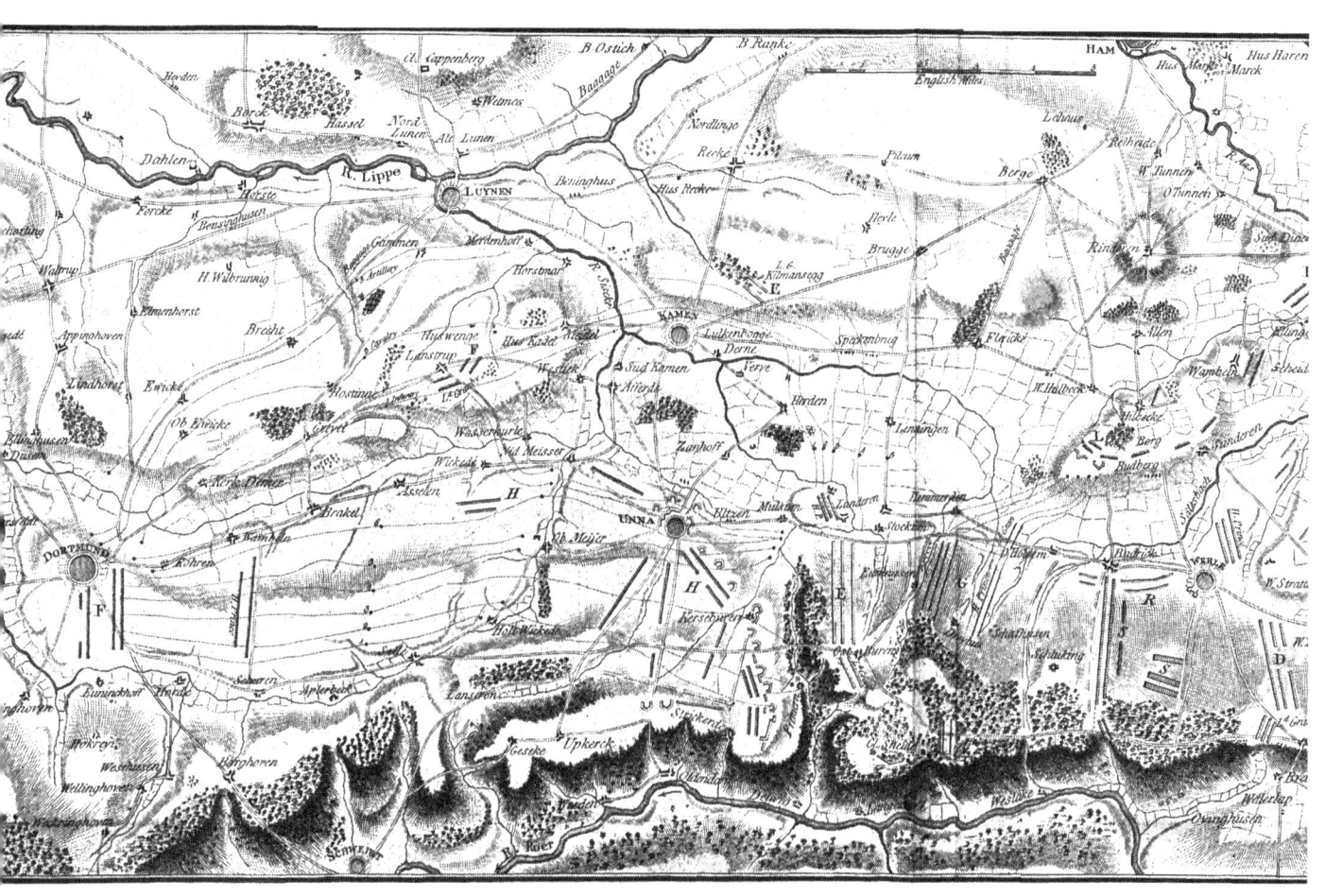

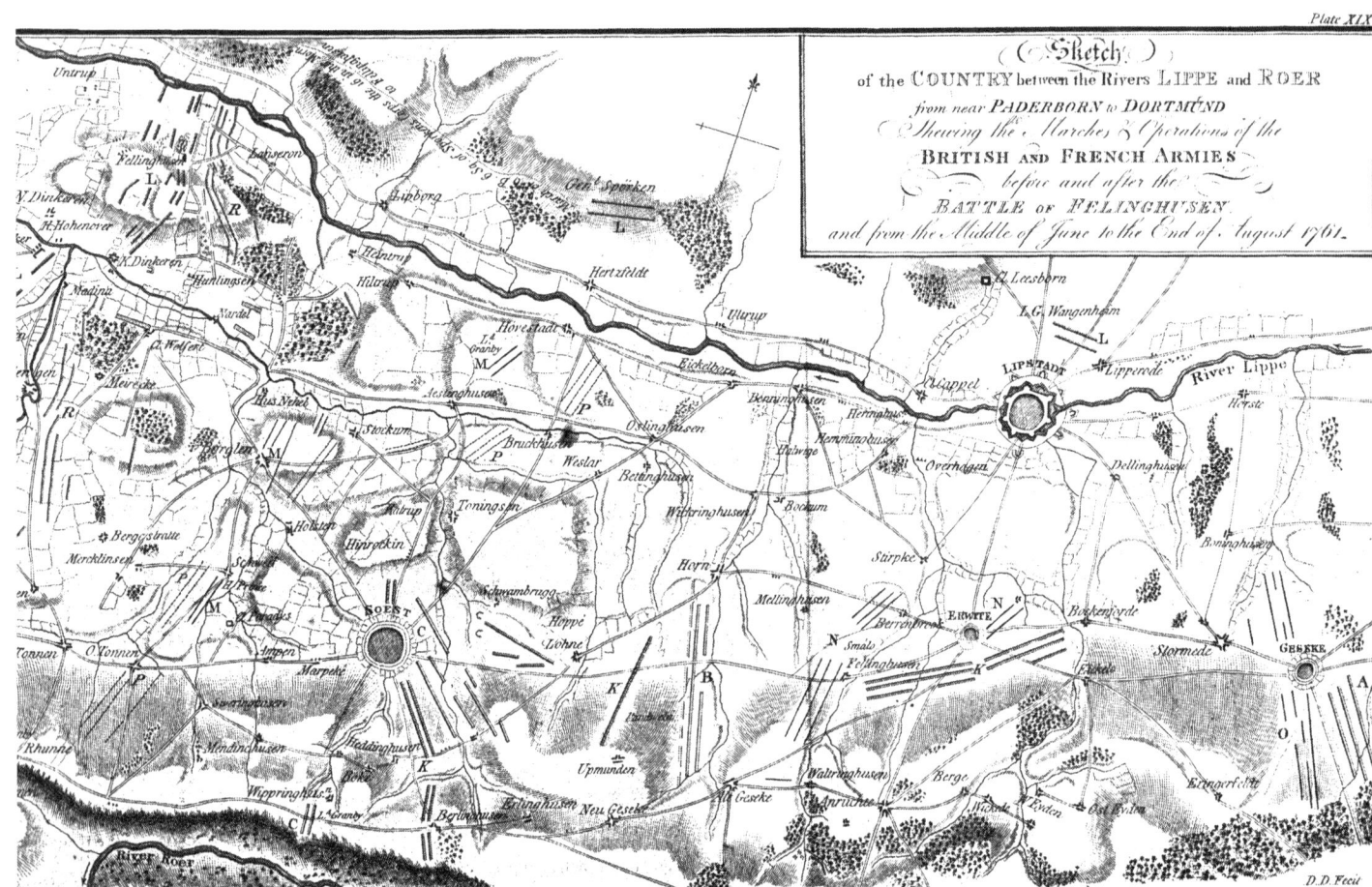

Plate XX

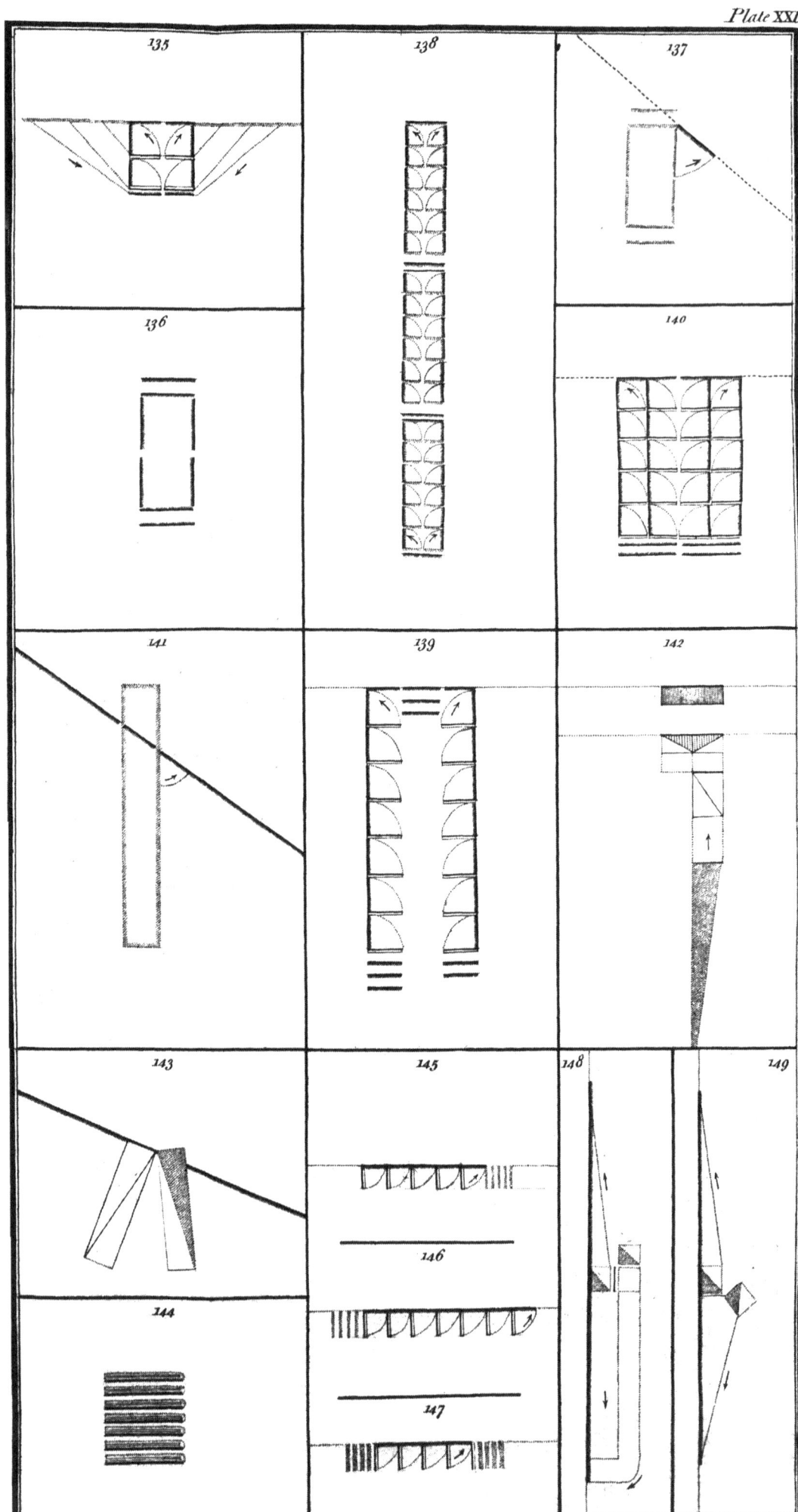

Plate XXII

151  158  152

154  155  157

156

The Positions are in general Represented { The First by — Yellow
The intermediate Ones by { Blue
Buff
Green }
The Last by — Red }

The dotted Lines and Arrows express the Direction of the Movements.

S. Neele Sculp.t 352, Strand.

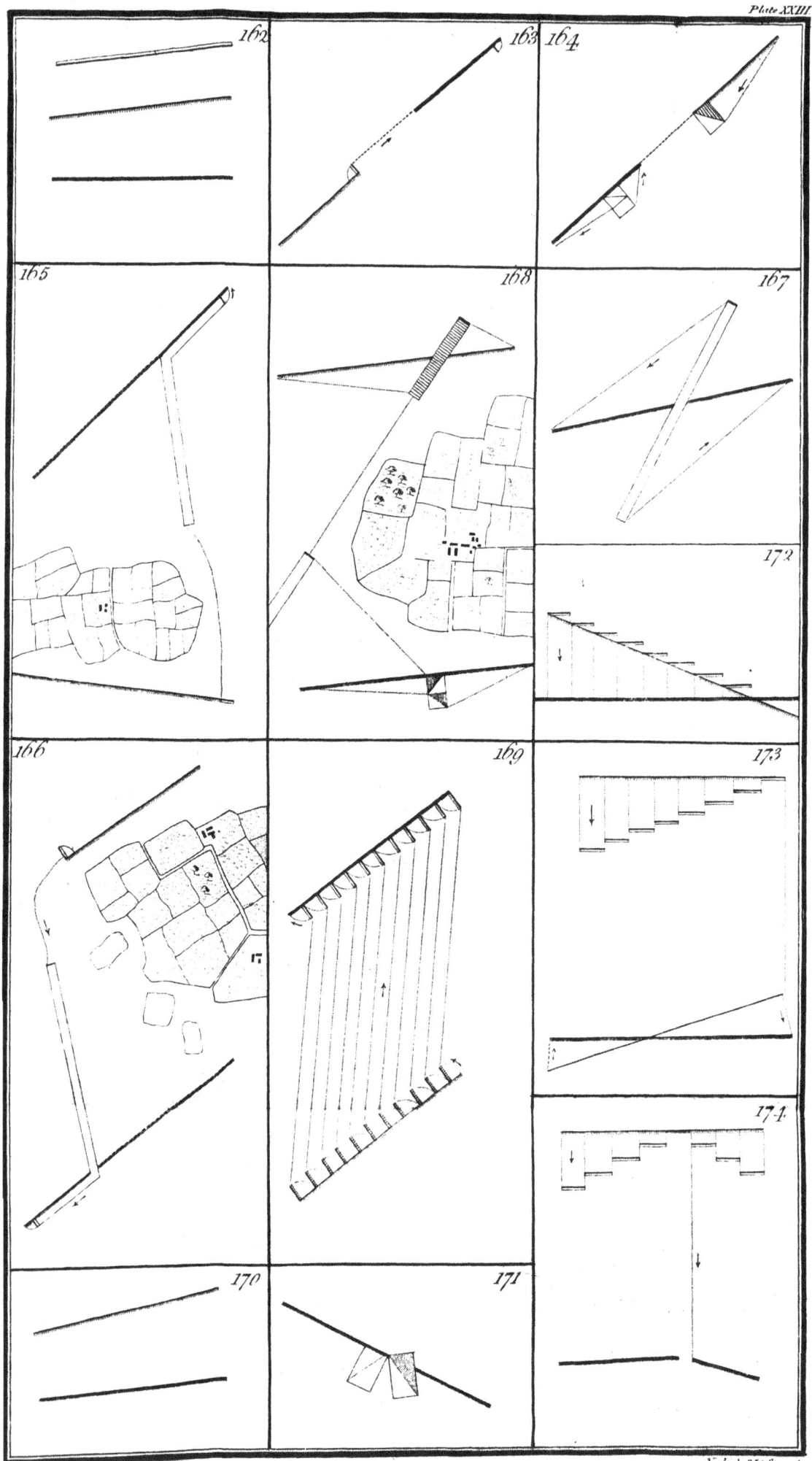

Plate XXIV.

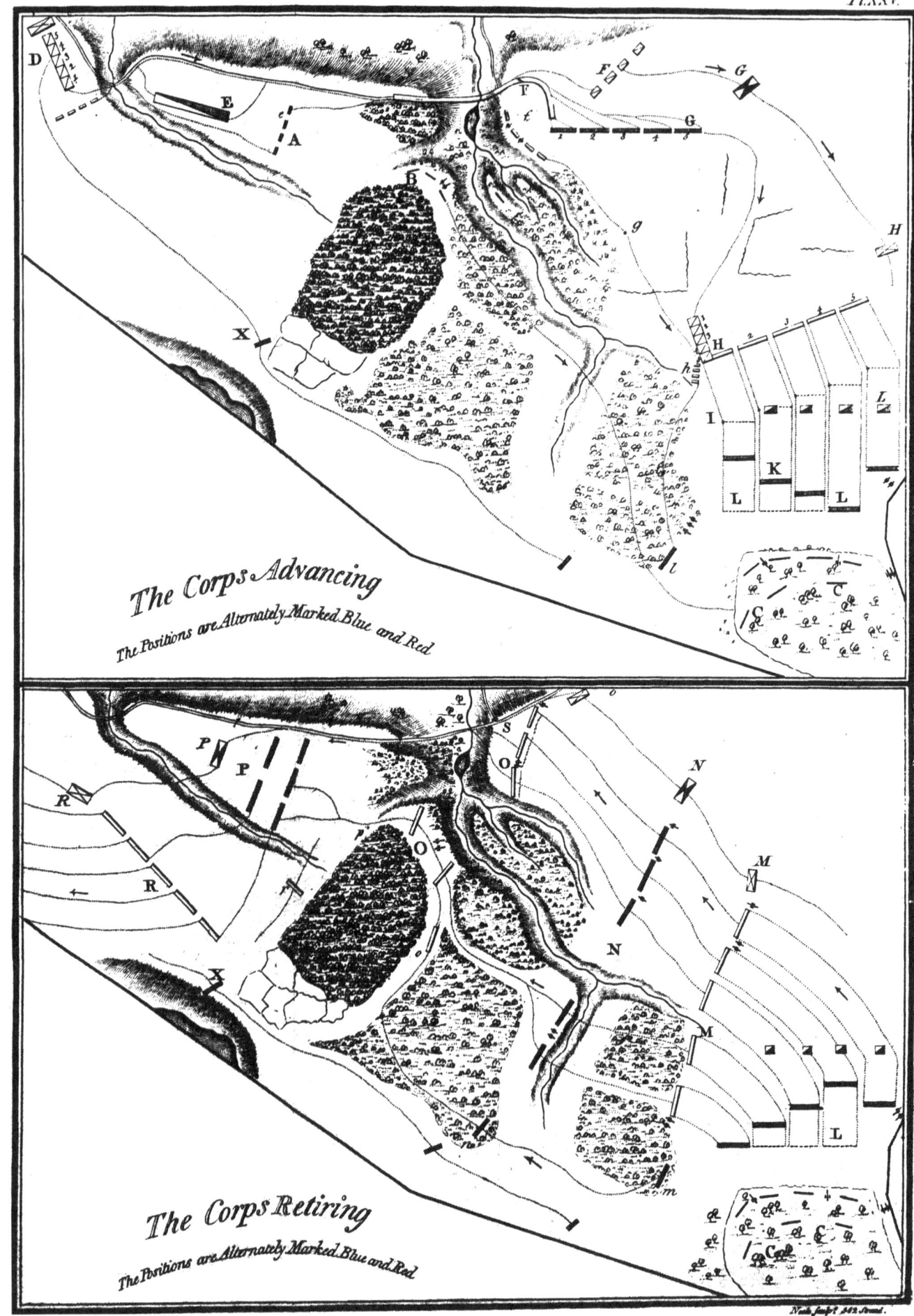

www.ingramcontent.com/pod-product-compliance
Lightning Source LLC
Chambersburg PA
CBHW080835010526
44114CB00017B/2313